Visual Ergonomics in the Workplace

Visual Egonomics in the Workplace

Jeffrey Anshel

UK Taylor & Francis Ltd, 1 Gunpowder Square, London EC4A 3DE
USA Taylor & Francis Inc., 1900 Frost Road, Suite 101, Bristol, PA 19007

© Jeffrey Anshel, 1998

All rights reserved. No part of this publication may be reproduced, stored in a retrieval system, or transmitted, in any form or by any means, electronic, electrostatic, magnetic tape, mechanical, photocopying, recording or otherwise, without the prior permission of the copyright owner and the publisher.

British Library Cataloguing in Publication Data
A Catalogue record for this book is available from the British Library.

ISBN 0-7484-0658-1

Library of Congress Cataloging-in-Publication Data are available

Cover design by Amanda Barragry

Printed in Great Britain by Athenæum Press Ltd, Gateshead.

Contents

Acknowledgements   vii
Introduction   ix

Chapter 1. Windows to the world   1
    Our ancestral eyes   1
    The Information Age   2
    An eye on the future   3

Chapter 2. The eye and visual system   5
    Eyeball basics   5
    How the eye works   6
    Refractive errors   7
    Seeing clearly now   10
    Binocular vision   10
    Visual skills for VDT users   11
    The development of myopia   12

Chapter 3. Your workspace and your eyes   15
    Defining visual ergonomics   15
    Visual stress at your job   17
    Viewing distances and angles   17
    Body posture and vision   19
    Office lighting   20

Chapter 4. Visual perception and VDTs   23
    Introduction   23
    Of CRTs and LCDs   23
    Display screen variables   26
    Purchasing a monitor   31

Chapter 5. Computer Vision Syndrome   37
    Introduction   37
    Eyestrain   38
    Headaches   38
    Blurred vision   39
    Dry and irritated eyes   40
    Neck and/or backache   42
    Light sensitivity   44
    Double vision   45
    After-images and color distortion   45

Chapter 6. Vision examinations   49
    Introduction   49
    The traditional eye examination   50
    The occupational vision examination   51
    VDT and vision questionnaire   51
    Vision screenings   52

Chapter 7.	**Vision in industry**	55
	Introduction	55
	The cost of eye injuries	55
	Industrial vision requirements	56
	Industrial eye protection	58
	Contact lens concerns	60
Chapter 8.	**Computing for the visually impaired**	63
	Introduction	63
	Working with visually impaired employees	64
	How blind persons use computers	65
	Low vision eye wear	67
	The Internet	67
	Resources	68
Chapter 9.	**Remedies**	69
	Introduction	69
	Reflections and glare	69
	Workplace conditions	72
	Brightness, contrast and color	78
	Work habits	79
	Computer glasses	80
	Eye exercises	84
Chapter 10.	**General eye care tips**	93
	Introduction	93
	Stress and vision	93
	Eye health concerns	95
	Aging factors	96
	Computers and contacts	97
	Vitamins and VDTs	98
	General body exercises	99
Chapter 11.	**The economics of visual ergonomics**	101
	Introduction	101
	Cost justification for ergonomics	101
	Providing eye care to employees	103
Chapter 12.	**Ergonomic standards**	107
	Introduction	107
	The Americans with Disabilities Act	107
	Technical standards	110
	The current state of ergonomic legislation	112
	The EC directives	114
Chapter 13.	**Epilogue**	117
Appendix A	*VDT workplace questionnaire*	119
Appendix B	*Occupational vision requirements questionnaire*	121
Appendix C	*Resources for the blind and visually impaired*	123
Appendix D	*Computer access products for blind and visually impaired users*	131
Appendix E	*Anti-glare screens*	137
Appendix F	*Additional resources*	139
Appendix G	*California Ergonomic Standard*	141
Glossary		143
Index		147

Acknowledgements

There are many people who have done research and investigations into the interaction of vision and computers. To all of these pioneers, I give my thanks.

There is one person, however, who stands above the rest in his commitment to research and teaching the principles which we are seeking. Dr. James E. Sheedy, O.D., Ph.D. is that person who has selflessly offered his knowledge and compassion to these ideas. He has attempted to organize this subspecialty into the mainstream of our profession and has given credibility to this area of study. His extensive research has been the foundation of many of our current principles of understanding. With his continued education of those who would follow in his footsteps, he has assured the computer-using public of a base of eye-care professionals who can successfully address their particular problems.

Introduction

Vision is our premier connection with the world. We use our eyes to interact with our environment in more than one million ways every second. The eye is an extension of the brain and is our direct link between our physical (outer) environment and our psychological (inner) mind. Over 80% of our learning comes from our vision and we most often 'believe' something when we can see it. Vision begins with 'visible' light – a portion of the radiation spectrum that simulates the nerve endings in the retina. The eyes can sense about ten million gradations of light and seven million different shades of color. The retina, which captures light and transforms it into nerve impulses, can form, dissolve and create a new image every tenth of a second.

The eyes are truly the windows to our world and to our soul. Because of their connection to the brain, they have influence over most of our cognitive thought processes. When the computer was first developed in the late 1940s, it was most often compared to the brain – with a network of interconnections and a sense of the logical thinking process. Even today, we tend to think that the computer is highly sophisticated and able to react quicker and more accurately than the human brain.

The comparison of the computer to the eye is a natural one. The computer generates and organizes information for our needs and we 'capture' that information with our eyes. You might think of the eyes as the connection between the two 'brains' we use. In order to maintain that link, our visual system might make certain adaptations to ease the flow of information. These adaptations can often lead to other complications. Therefore, care should be taken to assure that our computer viewing habits, viewing environment and visual condition are all considered wherever we use our eyes.

Organization of this book

This book is designed to give you a working knowledge of the eye and visual system so that you can make informative workplace decisions. Chapter 1 starts with an historical perspective on how our vision and visual system has had to evolve to keep up with our social development. Chapter 2 describes the working of the eyes and visual system, simplified for basic understanding for the non-medical reader. In Chapter 3, the interaction of the eye in the workplace is discussed, with emphasis on the functional vision of the worker. In Chapter 4, the technology of the VDT monitor and how the images are perceived by the eye are discussed.

Computer Vision Syndrome is a compilation of symptoms which arise from extended viewing of the VDT when the demands of the task exceed the abilities of the viewer. We discuss the various symptoms in Chapter 5 and address the possible reasons why they arise. How the eyes and visual system is tested and the role the environment plays in the examination procedure is discussed in Chapter 6. Although the main thrust of this book is the area of office ergonomics related to VDT viewing, we will also review the area of industrial vision and its many ramifications for eye safety and comfort in Chapter 7.

We often take our eyes for granted, often not having examinations for years at a time if we feel nothing is 'wrong'. There are many things that can go wrong and not everyone is fortunate enough to have 'perfect eyes'. The computer is also a useful tool for those who are visually impaired and there are many options available for them, which are discussed in Chapter 8.

The remedies for various problems are covered in Chapter 9. These include visual, environmental and physical areas which can be altered to increase VDT viewing comfort. There are many factors related to general eye care which can assist the reader in maintaining good visual habits. Chapter 10 discusses these factors and how to address them for the worker.

Every workplace has a certain budget within which to work. The changes which are suggested would be irrelevant if the cost considerations were not added to the equation. Therefore, the concern of the economics of ergonomics is addressed in Chapter 11. And then there are the governmental standards which also affect every workplace. We will attempt to deliver, in Chapter 12, the latest information which has passed as legislation or as standards in various countries. This area is obviously in a constant state of flux and should be viewed only as a general guideline.

It is hoped that this book serves as a cornerstone for a good understanding of the role vision plays in the lives of employees in the workplace. We spend almost one third of our lives in this environment and our eyes are our most important sense. The efficiency with which we see relates directly to how effectively and safely we perform on the job.

Chapter 1

Windows to the world

The eyes are simple tools. They are designed simply to catch light. However, the method by which they gather, filter and guide the light, as well as how our brain processes the information received by the eye, makes for the wonder of vision. In our interaction with our environment, there is little to compare with the contribution the eyes play. How we use our eyes and visual system dictates how well we survive in our environment. In order to have a sense of how the eyes and visual system are designed to work, we might first look at an historical perspective of our development – both physically and socially.

Our ancestral eyes

The current version of the species of man, *Homo Sapiens*, first appeared about 40,000 years ago. Our ancestors were designed to survive in a difficult and challenging environment. Finding food, shelter and protection were the first priorities, while the 'comforts' of clothing and amenities were secondary. Their bodies developed to support their needs: flexibility, agility and strength. And, if successful, they lived to the 'ripe' old age of between 25–30 years old, according to most authorities.

Likewise, the eyes of our ancestors were designed for similar types of survival. They were situated in a frontal position so the visual fields could overlap and work together, creating the sense of stereoscopic vision – the ability to perceive depth. They are near the top of the body in order to afford the longest range of seeing. Since they were being used mostly during the daytime, our 'bright' vision was keen; our night vision was adequate but secondary. The eyes were also developed for motility – moving in many directions, in conjunction with head movements, to be able to view a wide radius of the horizon.

This 'hunter/gatherer' type of visual system is the same one we are using today. We now, however, view our world for about 16 hours a day, with much of that in artificial light and in an environment which is mostly within arm's reach. We read small items of text in varying lighting situations for hours on end and struggle to meet deadlines which are imposed on us. During much of the year, our eyes are exposed to very little daylight for an extended period of time. We are asking our eyes and visual system to adapt to these adverse conditions and they must make the necessary adaptations to assure our 'survival'. These changes are slow to develop and not always successful for what we need to accomplish. Often sacrifices are made in one area of vision for the sake of seeing better in other situations.

The Information Age

Our success as humans developed mostly because of our ability to provide for ourselves. We learned how to hunt and to outsmart our predators. We also learned how to plant food and combine our hunting with agricultural skills so as to proliferate. Until the late 19th century, we were mostly an agricultural society who used these skills for survival. Our eyes maintained their ability to see clearly at distances during daylight hours with occasional near viewing.

The transformation in our visual development began in the late 1870s with the invention of the light bulb. With this new technology we were now able to extend our comfortable viewing into the night-time hours. Other developments and technological advances lead to the start of the 'Industrial Age', where machines were developed to help us in our daily living. One prime example is the automobile, which made an obviously major impact on our society. Driving vision consists of a combination of distance and intermediate viewing, with the distance being most critical. But along with this development was a need for closer inspection of machine parts, dials and other mechanical devices. Near point viewing was a critical task and the success of these and other machines depended on it. The success of employees working in this situation was directly related to how efficiently they used their eyes. Poor vision reduces performance and productivity of workers as well as increasing their risk of having an accident on the job.

A significant part of this development was a parallel development of the electronics field which assisted in the functioning of the machines. About midway through the 20th century, the computer was developed and our society once again began to shift. Because of its unique ability to do basic arithmetic and other numerical calculations, it gradually became (and continues to become) an integral part of our newly dawning 'Information Age'. Although it wasn't obvious at first, the visual requirements for computing are, once again, different from what was required for viewing other types of mechanical equipment. Working with a self-illuminated video display terminal (VDT) screen at an awkward angle of view and unique working distance is yet another adaptation our visual system must make. It is noteworthy that this last transformation occurred less than 100 years from the previous shift, which is extremely rapid in historical terms.

An eye on the future

As computer viewing gets more commonplace and becomes our standard of operation, the problems which arise are bound to be compounded. While only 10% of the workforce was using computers for their daily activities in 1980, that percentage has climbed to over 50% today and is expected to reach the 100 million mark by the end of the century. The computers are getting faster and 'smarter', the technology of monitor design and quality is improving and they are getting more affordable. Yet, there are still major considerations to address. In 1991, a Lou Harris poll cited computer-related eyestrain as the number one office health complaint in the United States. In 1994, a study by NIOSH indicated that 88% of the 66 million people who work at computers for more than three hours a day, nearly 60 million, were suffering from symptoms of eyestrain. And no matter what type of input device is found to make our entries easier (e.g., voice activation) the output of computers will always involve the visual system. Although problems with computer eyestrain outstripped carpal tunnel syndrome and other more highly profiled office health complaints more than half a decade ago, the size of this epidemic has not received the attention it warrants. The reason is that a key cause of computer eyestrain has not been well understood.

An additional concern is the potential for vision and other physical problems in our children, who are growing up using this new technology on a full-time basis. As it appears today, very few considerations are being addressed for the computer-using habits of children. Ergonomic furniture, monitor placement, mouse dimensions, proper lighting, body posturing and many other factors have yet to be considered. Add to this equation the amount of time children spend looking at computer-generated games and school work and the potential for serious complications is increasing.

In addition, the role of eye care practitioners will also shift as the viewing situations of their patients change. The role of eye care providers must keep up with the demands of their patients. The doctors must analyze problems that arise due to the interaction of workers/patients with their environment, be called on to design optimal viewing environments, and evaluate those environments to improve visual performance. This may well transform the way routine eye examinations are performed.

Our 'future vision' may well be one of adaptation, once again. However, several considerations must be addressed and, as of this writing, it appears that we don't even have all of the questions, much less the answers, that we require. It is likely that just as we now routinely accept that sunglasses are appropriate for a sunny day, we may need to accept that computer viewing glasses are just as appropriate for our VDT viewing. Our eyes will need to learn to adapt to our new viewing situations, but we must know how we can help ease that transition.

Chapter 2
The eye and visual system

Eyeball basics

This book is not intended as a technical medical synopsis of the anatomy and physiology of the human visual system. However, you should know some basic information about how the eye is put together and how it works in order to have some background with which to make intelligent decisions regarding vision requirements in the workplace. We'll start with some basic eye anatomy just so you know what is what and where it is. Then you can understand how the parts work together to create this fascinating organ.

The eyeball is basically that – a ball. Its diameter is roughly an inch, and it's about three inches in circumference. The part of the eye that is visible to the world – between the eyelids – is actually only one-sixth of the eye's total surface area. The remaining five-sixths of the eyeball are hidden behind the eyelids. The outer surface of the eye is divided into two parts: the *sclera* (SKLER-ah), the white part that is the outer covering of the eye, and the *cornea* (KOR-nee-ah), the transparent membrane in front of the eye. The cornea, which is steeper in curvature than the sclera, may be difficult to see because it is transparent and the colored *iris* (EYE-ris) is behind it. You can see the cornea most easily if you look at a friend's eye from the side. The sclera is made of tough fibers, which allow it to perform its function as the supporting structure for the contents of the eyeball. It has a white

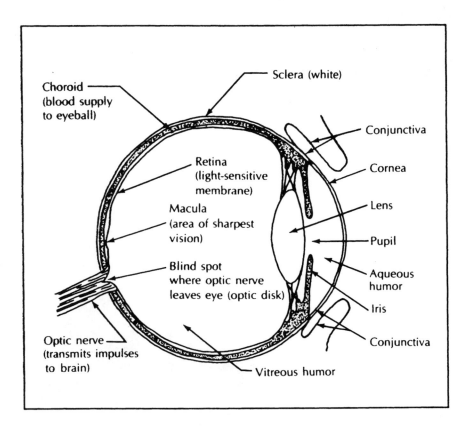

Figure 2.1 Side view of an eyeball.

appearance because the fibers are light in color and because there are very few blood vessels in it.

Just inside the sclera and covering all of the same area is the *choroid* (KOH-royd), which is the main blood supply to the inner eyeball. And just inside the choroid is the *retina* (RET-in-ah), the nerve membrane which receives the light. In addition to the blood vessels of the choroid, there are also blood vessels that enter the eye through the optic nerve and lie on the front surface of the retina. They supply nutrition to the retina and other structures inside the eye. These parts all seem to be very basic when you think of what an eye must do. The eye needs protection and support (provided by the sclera), a blood supply (provided by the choroid and through the optic nerve) and a mechanism for seeing (provided by the retina).

As you look at an eye, the first thing you'll notice is the colored iris. If you look closely at an eye, you'll see that the iris is actually enclosed in what's known as a chamber, which is a medical term for a closed space. The iris is also surrounded by a watery fluid called the *aqueous* (AY-kwee-us) humor or aqueous fluid. 'Humor' doesn't have anything to do with being funny, it's just the Latin word for fluid. Just behind the iris is the lens, which provides the focusing part of the vision process. The lens is transparent and can't really be seen from the outside unless special equipment is used. Behind it, and filling the main chamber within the eye, is the *vitreous* (VIT-ree-us) humor. The vitreous humor is more gel-like and less watery than the aqueous humor and helps in the support of the retina and other structures.

How the eye works

Let's look at the visual process by starting at the beginning. Light enters the eye by passing through the cornea, the aqueous humor

The eye and visual system

and the pupil; is focused by the lens; and then goes through the vitreous humor and onto the retina. It was noted that the retina is actually an extension of the brain. That's right! The nerve fibers from the retina go directly into the brain.

The light that strikes the retina first stimulates chemical changes in the light-sensitive cells of the retina, known as the *photoreceptors*. There are actually two kinds of photoreceptors: the rods, which are long, slender cells, respond to light or dark stimuli and are important to our night vision; the cones, which are cone-shaped, respond to color stimuli and therefore are also called color receptors. There are about 17 times as many rods as there are cones – about 120 million rods and 7 million cones in the retina of each eye. These rods and cones interconnect and converge to form a network of about one million nerve fibers that make up *each* optic nerve.

When light strikes the rods and cones, they convert the light energy to nerve energy; we'll call this nerve energy a visual impulse. This impulse travels out of the eye into the brain via the optic nerve at a speed of 423 miles per hour. It first reaches the middle of the brain where a pair of 'relay stations' combine the visual information it is carrying with other sensory information. The impulse then travels to the very back part of the brain, the visual cortex. It is here that the brain interprets the shapes of objects and the spatial organization of a scene and recognizes visual patterns as belonging to a known object – for example, it recognizes that a flower is a flower. Further visual processing is done at the sides of the brain, known as the temporal lobes. Once the brain has interpreted vital information about something the eyes have 'seen', it instantaneously transfers this information to many areas of the brain. For example, if the information is that a car is moving toward you, it is relayed to the motor cortex, which is the area that controls movement and enables you to get out of the way. This is located in a band that goes over the top of your head from just above one ear to just above the other ear.

So, vision is really the combination of the eyeball receiving the light and the brain interpreting the signals from the eye.

Refractive errors

The process I've just described is how the normal eye and visual system functions when it works perfectly well. This condition of the optically normal eye is called *emmetropia* (em-e-TROH-pee-ah). This isn't, unfortunately, always the case. Very often there is something that goes wrong and the visual process is disrupted. About 50% of the adults in the United States have difficulty seeing clearly at distance and about 60% have difficulty seeing at near with no corrective lenses. One of the more common problems is the mis-focusing of the light as it is directed onto the retina. The light can focus too soon, too late or be distorted. Because the bending of light is technically called 'refraction', the mis-focusing of light in the eye is called a 'refractive error'.

First, let's try to get these terms straightened out. Near-sightedness, also called myopia (my-OH-pee-ah), means having good near vision but poor distance vision. For the myopic person, a distant image (an image at least twenty feet away, so that the eye's lens is as relaxed as it can be) falls in front of the retina and looks blurred. Nearsightedness results when an eye is too long, when the

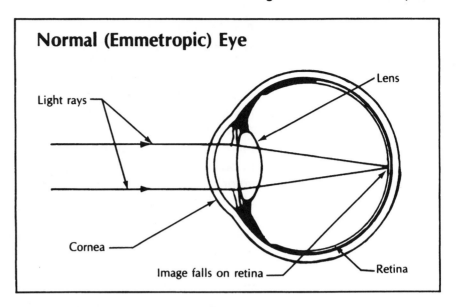

Figure 2.2 In the normal eye, an image falls exactly on the retina. The shape of the eyeball and cornea are normal, and the eye's lens has normal flexibility and focusing ability.

cornea is too steeply curved, when the eye's lens is unable to relax enough to provide accurate distance vision, or from some combination of these and other factors.

Farsightedness, also called hyperopia (hy-per-OH-pee-ah), is *not* exactly the opposite of myopia. For the hyperopic person, an object that is twenty feet or more away (so that the lens is relaxed) is directed past the retina, so that it looks blurred because it hasn't yet focused. Farsightedness results when an eye is too short or the cornea too flat, or from some combination of these and other factors. The main difference between these two conditions is that the eye can increase its focal power (to some degree) to compensate for

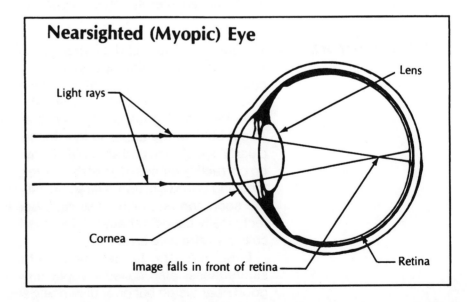

Figure 2.3 In the nearsighted eye, the image falls in front of the retina when the lens is in its relaxed state, viewing an object that is at least 20 feet away. The image is blurred.

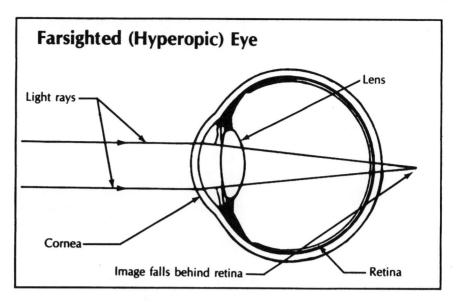

Figure 2.4 In the farsighted eye, the image falls behind the retina when the lens is in its relaxed state, viewing an object that is at least 20 feet away. The image is blurred.

farsightedness, whereas it can't reduce its power to compensate for nearsightedness.

Theoretically, the surface of the cornea should be almost spherical in shape, like the surface of a ball, so that when light passes through it, it can be focused at a single point. However, nature isn't always perfect, and the cornea is often 'warped' so that it more closely resembles a barrel than a ball. The lens too can be irregular in shape. These distortions can be significant enough so that the light that passes through the cornea and lens in the vertical orientation will focus at a different spot from the light that passes through in the horizontal orientation. Now you have two points of focus with a blur between them. This is known as *astigmatism* (a-STIG-ma-tism).

If the difference between these two points of focus is great enough, the eye will strain trying to decide which point of focus it should use. You might then develop occasional blurring of vision, tiring or possibly headaches. Astigmatism in small amounts is very common and not of great concern. However, about 23 million Americans have a significant amount of astigmatism which requires correction.

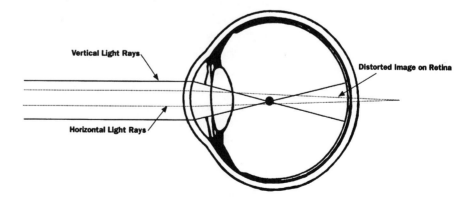

Figure 2.5 In astigmatism, the image that enters the eye is distorted (usually by the cornea) and does not have a single point of focus.

Glasses correct astigmatism by having curvatures that compensate for the curvature of the eye. This is a simple optical correction, and the glasses will not change the amount of astigmatism; that is, they won't 'cure' the problem.

Seeing clearly now

Recall the last time you visited your eye doctor's office. You probably got a full examination, had what seemed like a hundred different tests, and you asked: 'How are my eyes?' The answer could have been: 'You have 20/20 vision!' You then walked out of the office satisfied that your eyes were in good shape. But are they? What does 20/20 refer to and what does it mean?

The numbers are really just a notation that relates to the resolving power of the eye. Resolving power means how sharp your sight is, which we can define as your ability to distinguish two points from each other and not see them as just one point. If your vision is 20/20, it means that you're seeing at twenty feet what the 'optically normal' eye can see at twenty feet – that is, that your eyes can distinguish one point from another on a specific line from a standard eye chart placed twenty feet away. The chart is called a Snellen chart. If your vision is, let's say, 20/40, it means that you can see at twenty feet what the normal eye can see at forty feet (you have to be closer). And, if your vision is 20/100, you must be at twenty feet while the normal eye can be a hundred feet away and see the same thing as clearly. In short, the larger the bottom number is, the poorer your resolving power, which is also known as your visual acuity. Visual acuity is measured for distance and near vision. So now you know that 20/20 is something like a grading of eyesight.

Binocular vision

Seeing a clear 20/20 is certainly a good indication that your eyes might be doing a good job. However, sharp eyesight is just one of the functions that your eyes perform. Since we have two eyes, we must make sure that they are working in harmony with each other. One of the most fascinating abilities of the visual system is to take images from two eyes and put them together into just one picture. You don't normally see two images, so the idea might sound strange, but double vision can occur and is one of the most dangerous manifestations of vision problems. Imagine seeing two cars coming at you as you drive down the road!

Here's how the brain keeps us from going off the road. Let's assume that you have two eyes and they are both working equally well. As you look at just one object, each eye receives an image of that object. Both these images are transmitted back to your brain, but they are then 'fused' together by the brain into one image. In order for that to happen, both eyes must be pointed at the same object, and the images have to be approximately equal in size and clarity.

Now, if one eye does not aim at the same spot as the other, each eye will be 'looking' at a different object, and the two images won't match up. When the images are transmitted back to your brain, they will stimulate two different groups of brain cells, and you will experience two images: seeing double. After a short time, your brain will decide to turn off, or suppress, the picture from the eye that is pointed in the wrong direction so that you can see one image again.

This suppression is necessary for our visual survival, but it is not the way we were made to see.

This suppression of an image is the brain's way to make our daily tasks easy and comfortable in stressful situations. Thus, you might think that suppression of an image is devastating but it actually works pretty well. What is more serious, however, is when there is a 'competition' between the eyes and they are struggling to work together. This problem is much more common than the suppression problem. It is this competition that causes the person to struggle with reading tasks and can lead to poor reading comprehension and job performance. Adequate binocular function is important to successful work related tasks.

Visual skills for VDT users

Let's take look at the VDT viewing task and note the various visual skills that are required to view it comfortably and efficiently.

First, there is visual acuity. If the letters viewed on the screen are not clear, then the user cannot perform the task effectively. Blurred letters at the VDT distance can be caused by a number of conditions, which we will discuss shortly. For now, just realize that clarity, without excess effort, is essential to proper VDT use. Next, there is *accommodation*. Accommodation is the act of refocusing the light when viewing a near object as opposed to viewing a distant object. This entails using the crystalline lens (just behind the iris) to adjust its shape, which in turn alters its focusing power. There are many facets to evaluating accommodation but the two of significance here are maximum focusing ability and the flexibility. The maximum focusing ability determines if the image can be brought into clear focus onto the retina; the flexibility determines if it can be done easily and also reversed to regain distant vision with ease. Both of these factors must be mastered by the visual system of the VDT user to assure clear and comfortable viewing.

Binocular vision – the efficient use of both eyes together – is also necessary but this might be subject to debate. It is true that many 'one-eyed' people can use a VDT with ease and comfort, so you might think that binocular vision is not important. However, remember that the suppression of one of the eyes' images is effective for our daily function but is not optimum. It is, again, the close competition between the two eyes which can lead to decreased performance if not accomplished with efficiency. Scanning and tracking, both eye movement skills, must also be performed smoothly for the user to make accurate visual moves. Scanning involves the jumping of the eyes over the image as a whole with the ability to perceive many images at once and then locate the one desired. Tracking is the type of movement which is used in reading, where small jumps are made at regular intervals so as to follow an orderly progression. The VDT user must be able to perform these skills efficiently to keep from over using and tiring the eye muscles.

Glare recovery is a significant factor in VDT use because of the many sources of glare found in the workplace. This topic will be covered in much more detail later but the eyes' ability to recover from glare is normally determined by the function of the retina and other optical properties within the eye. Proper nutrition and eye health are critical factors in glare recovery abilities. Hand–eye coordination is also a function which the user must master due to the input methods involved with computer use. Using the feedback from the visual system, keyboard entry will determine what the next item is to be. Also, efficient mouse manipulation is determined by hand-eye coordination abilities.

More details of visual requirements for VDT use are discussed in Chapter 7.

The development of myopia

The latest demographic figures show that 58% of the US population as a whole wears some form of vision correction. About 32% of them are nearsighted. This should be a rather surprising number considering that less than 2% of the population is born nearsighted. There is still some controversy as to what is the exact cause of myopia, but new research is shedding some light on the subject (Zadnick *et al.*, 1995).

It's been believed for a long time now that myopia is inherited, and we shouldn't overlook the contribution of a heredity factor. But it's probably not the whole story because myopia is much more prevalent in people and societies where close work is a significant part of daily life. Studies have found, for example, that myopia is almost non-existent in uneducated societies (such as early Eskimos or some African tribes) and that myopia increases in proportion to the amount of education in any given society (Young *et al.*, 1969). In other words, the more reading and near-point work the society does, the higher the incidence of myopia.

In a similar vein, studies have been conducted with Navy submariners (Schwartz and Sandberg, 1954), who are submerged for months at a time, in a space where their maximum viewing distance is about eight feet. The studies showed an increase in myopia during these extended periods of confinement.

Dr. Francis Young of Washington University has done similar research with Rhesus monkeys (Young, 1963). He kept the monkeys in confined areas during various developmental periods of their early life. His research showed that the shorter the maximum viewing distance and the longer the confinement, the more myopia the monkeys developed.

So what does this say about the way our eyes develop? As with any biological system, our visual system will change in response to stress. While reading, your eyes are focused at a close distance – usually about 14 to 16 inches. The eye accomplishes this focusing through the process of accommodation. If this posture is maintained for long periods of time without a rest, the eyes slowly adapt to the position in order to reduce the stress on the muscles controlling each eye's lens. Once adapted, the eyes can see more clearly close to with less effort. It's as if the muscles get comfortably 'stuck' in the near-focus position. To make matters worse, when the eye muscles work constantly to accommodate for near work, they cause some increased pressure to build up in the eye. Eventually, this pressure causes the eyeball itself to lengthen (which relieves the pressure), moving the retina farther back from the lens than it was originally.

So, what is the result of all this stress at the close working distance? Myopia. When the myopic eye relaxes the accommodation effort and attempts to refocus for distance, the image is blurred because it is too far forward of the retina. This process doesn't happen by just reading steadily for a night or two. It's a gradual adaptation that your eyes go through as they react to the strain of overwork.

As you might expect, myopia increases in children as they spend more and more of their time focused for near-vision activities. According to early statistical studies (Hirsch, 1952), about 1.6% of children entering school in the United States have some degree of myopia. That figure grows to 4.4% for seven- and eight-year-olds, 8.7% for nine- and ten-year-olds, 12.5% for eleven- and twelve-year-olds and 14.3% for thirteen- and fourteen-year-olds. We used to say that the progression (worsening) of myopia stabilized at maturity, about 21 years of age. However, over the past 15 or so years, eye care professionals have seen more myopia progressing well into the late twenties or even thirties. The reason? We're not quite sure, but computers and their accompanying VDTs are almost certainly to be considered one of the culprits.

References

Hirsch, M.J. (1952) Visual Anomalies among Children of Grammar School Age, Archives of the American Academy of Optometry, 23:11.

Schwartz, J. and Sandberg, N.E. (1954) Visual Characteristics of Submarine Population, Medical Research Laboratory, US Navy Sub Base, North London 13.

Young, F.A., Leary, G.A., Baldwin, W.B., West, D.C., Box, R.A., Harris, E. and Johnson, C. (1969) The Transmission of Refractive Errors Within Eskimo Families, Archives of the American Academy of Optometry 46:9.

Young, F.A. (1963) The Effect of Restricted Visual Space on Refractive Error on the Young Monkeys' Eyes. *Investigative Ophthalmology*, **2**.

Zadnick, K., Mutti, D., Adams, A.J. and Fusaro, R. (1995) Orinda Longitudinal Study of Myopia, Proceedings of the American Academy of Optometry, Dec.

Chapter 3

Your workspace and your eyes

Defining visual ergonomics

As you are reading this book, you are probably familiar with the term and concept of ergonomics. It can be defined as the concern with the design of a work system in which humans interact with machines. This interaction should address the fitting of the workplace to the worker, not the worker to the workplace. Two general categories would include industrial ergonomics and office ergonomics. Industrial ergonomics has been around for many years and covers industries such as aerospace, transportation, manufacturing, and other industrial settings. More recently, the area of office ergonomics has become a more popular concern since the introduction of the VDT into the workplace. Industrial visual concerns will be addressed in a later chapter.

The definition of ergonomics has recently been broadened to include: (1) equipment and workstation design; (2) job design; (3) work–group relations; and (4) education of employees and training staff. Management has discovered that employees perceive a direct relationship between job stress, VDT work and their health. This is a major step in the merging of ergonomics with other workplace programs to enhance employee performance. Studies (Sauter *et al.*, 1981) have shown dramatic productivity increases up to 20% resulting from ergonomic improvements in the design of the job and the workplace. A concomitant increase in worker satisfaction and a decrease in physical complaints were also reported.

Much of the attention surrounding office ergonomics has been with the upper extremities. These include Repetitive Motion Injuries (RMIs) such as carpal tunnel syndrome, tendonitis, bursitis, and other inflammations and conditions which can be attributed to repetitive tasks. These conditions should be properly categorized as Cumulative Trauma Disorders (CTDs). The common characteristics of all CTDs include: (1) the symptoms are work related and associated with cumulative trauma or repetition; (2) they are disorders of muscles, tendons, bones and nerves; (3) they are caused or aggravated by repeated exertions or movements; and (4) they require a long period of time to develop and to recover.

This book is directed toward the visual system and its function in the workplace. One might think that vision problems in the workplace are separate from the most common CTDs. However, it can be shown that computer-related eye and vision problems meet most of the requirements of CTDs. Cumulative Trauma Disorders are largely the result of using a part of the body which functions well under normal activity or work loads, but for which problems develop when it is taxed repeatedly and stressfully. Likewise, the visual problems that result from work at a computer are the result of stressing some aspect of the eye or visual system to the point that it causes symptoms. Most employees with vision and eye problems at a computer do not experience those symptoms when they perform less demanding visual tasks. If they have a visual disorder such as a focusing problem or eye muscle imbalance, for example, it may not produce symptoms under normal daily visual tasks. When they perform a demanding visual task such as display screen work, however, the disorder is expressed through symptoms. This is essentially the same way in which symptoms of CTDs express themselves (Sheedy, 1992).

The primary difference between computer-related eye problems and CTDs is that most eye-related symptoms are daily events, so that the worker begins anew each day without carry-over symptoms from the previous day. Such is not the case with classic CTDs in which the symptoms become chronic. In this respect, vision and eye symptoms fit more into the category of 'localized fatigue' (Armstrong, 1991). However, vision and eye symptoms don't fit perfectly into the localized fatigue definition either. In most instances of localized fatigue, the body is able to adapt to the conditions that cause the fatigue, so that the fatigue no longer occurs with regular exposure, such as a runner building up an endurance for a particular race distance. This result is common with muscular fatigue or cardiovascular fatigue but does not appear to occur with eye-related symptoms, in which there is no indication of resistance that is developed to eliminate the fatigue.

An example of this repetitive nature of visual tasks at the VDT involve the accommodative and convergence mechanisms of the visual system. They are asked to do a repetitive task of changing the eye posture at varying positions for extended periods of time. This often consists of copying text from hard copy to display screen – viewing the text, then the screen, then back to the text, and so on. If the distance from the two sources is extreme or the number of actions during a normal workday is large, the repetitive action will fatigue the eyes. Additionally, if the screen is not in a 'proper' position (which will be addressed shortly), the eyes must adapt an awkward posture. This often creates a misalignment of the convergence system of the eyes. Maintaining this posture will lead to visual stress and other eye conditions, such as dry eye syndrome.

Visual ergonomics consists of setting up a work space in which clear and comfortable vision can occur. Whether this includes a visual compensation, such as glasses or contact lenses, is dependent on the state of the visual system of the individual. It also includes the areas of body posture, office lighting, job stress and other factors which are discussed later in this chapter.

Visual stress at your job

Each occupation has its own unique working environment and visual demands. Table 3.1 lists many of the various VDT-related tasks and the amount of time spent using the display screen.

From this chart you'll notice that the more service-oriented jobs require more screen use and tend to be in more stressful positions. These are the jobs which should have their VDT use specifically addressed as a key issue in their working situation. It behoves the reader to take the time to evaluate the workers under his or her management with regard to the type of visual stress associated with their particular type of work.

Viewing distances and angles

The two most obvious differences between reading text on paper and a display screen involve the viewing distance and the angle of viewing. The literature is awash with many references to the 'optimum' viewing distance for VDT work (for example, Lehmann and Stier, 1961; Grandjean et al., 1983; Hill and Kroemer, 1986; Jaschinski-Kruza, 1988). Recommendations for optimum distance vary from 16–30 inches and for viewing angle from 5–60° below the horizontal line of sight, both of which offer quite a large variation from which to choose. However, these recommendations do not take into account the uniqueness of each person's own visual system, posture, and

Table 3.1 Various occupations have their own specific visual requirements, especially when it deals with VDT viewing

Visual Stress Using a VDT		
Task	*Usage*	*Viewing*
Data Entry	Heavy	Hard Copy
Customer Service	Heavy	Screen
Airline Reservation	Heavy	Screen
Word Processing	Heavy	Hard Copy/Screen
Editor / Author	Medium	Screen
Programmer	Heavy	Screen
CAD/CAM	Heavy	Screen
Executive	Low	Hard Copy
Air Traffic Control	Heavy	Screen
Accountant	Heavy	Hard Copy/Screen
Graphic Artist	Heavy	Screen
Manager	Low	Hard Copy/Screen

how the accommodation process reacts while viewing a display screen.

One would think that when gazing at a particular object (especially within arm's reach) that the eye would simply focus directly on the target. Visual research, however, has discovered a phenomenon called a 'lag of accommodation'. This lag can be thought of as a difference between the plane of the target being viewed and the point where the eye actually focuses. For example, while reading a book at 16 inches, the eyes often will *converge* to the 16-inch distance yet the accommodation will be at somewhere between 17–20 inches or so. The actual distance will vary with many factors, including refractive error, lighting conditions and psychological stress.

While viewing a display screen, the exact object of focus should be the posterior surface of the glass, where the image is created. However, with the lag of accommodation, the point of focus is often further back. A study done by Murch (1982) found that the eye cannot focus on the information displayed on a VDT screen with the same accuracy as the printed page. This practice places an extra effort on the accommodation system which is enhanced with the additional viewing of display screen, printed hard copy, and office environment. Rouse *et al.* (1994) found that the lag of accommodation was not only greater during VDT viewing but that it increased more with the length of time the text was being read. This further implicates the poor image of the display screen as a major factor in poor viewing conditions of computer users.

Another concept related to accommodation is called the Resting Point of Accommodation (RPA). This is the distance at which the eye focuses when there is no object on which to focus (dark field). At one time this resting point was thought to be infinity, but recent studies show this not to be the case (Owens, 1984). It was found that the RPA is actually a point within that 20-foot distance and it varies with the individual, much as the lag of accommodation does (Liebowitz, 1975). This resting point averages around 80 cm ($31\frac{1}{2}$ inches) (Kruegar, 1984). There seems to be a small but significant transient shift toward myopia following sustained near point work (Blustein *et al.*, 1993).

Theoretically, this means that viewing a VDT monitor any closer than this RPA distance contributes to eyestrain. However, there are many other factors to consider. For one, the relationship between the accommodative effort and the convergence between the eyes is

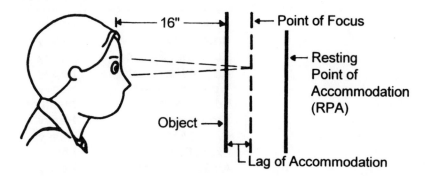

Figure 3.1 The 'lag of accommodation' is the point to which the eye actually focuses, usually behind the target being viewed.

Your workplace and your eyes

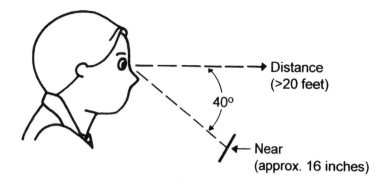

Figure 3.2 The normal reading posture requires approximately a forty degree downward gaze from the distance viewing direction.

strained, therefore creating one of the major causes of tiring of the eyes while viewing a display screen.

The angle at which you view your monitor is another critical factor. Most VDT users have their screens located directly in their straight-ahead line of sight, as when viewing at a distance object (often on top of their CPU). A Swedish study (Bergqvist *et al.*, 1994) found that having the VDT at eye level was directly associated with symptoms of eyestrain. Since the point of focus and convergence between the eyes is ideally between 20–28 inches, the eyes must turn in to that point. However, when they do, they do not accomplish it in the same horizontal plane as the straight-ahead position. As they converge closer to each other, the line of sight lowers, so that by the time they are viewing at 16 inches, the viewing plane is lowered to about 40–45° below horizontal. Since the display screen should be somewhere between the 20 feet and 16 inches (possibly 24 inches), the horizontal level should be lowered by somewhere between 10 to 20 degrees. This will be discussed in more detail in Chapter 9.

Body posture and vision

Those who study human kinetics are familiar with the concept of 'the eyes lead the body'. When there is an external stress on the eyes, the body will be re-positioned to reduce this stress. One example often used is that of a young child piano player who is giving his/her first recital. If they hit a sour note, the first reflex is to squint their eyes and lean forward toward the music to see it better. The reason for this reaction comes from the auditory feedback (the sour note) causing the eyes to over-focus, which in turn causes the body to lean forward to see the notes so that they, hopefully, become clearer.

Similarly, if the VDT users are having difficulty viewing their screen, their accommodative system will over-react, causing them to over-focus at the screen distance. To compensate, they must lean forward toward the screen in an attempt to clear the image. This will engage many body mechanisms, including the neck, shoulders and back to support this posture. Secondarily it will also put more stress on the wrist and fingers because of the angle of the arms. Thus, it appears that visual difficulties can lead to all sorts of problems for the VDT user.

There is a common notion today that the body should work in a 'neutral' position. This refers to a posture whereby there is no excessive strain or stress on the joints and they are essentially in a straight, or neutral, position. This may also be interpreted as adopting a static posture with little or no movement throughout the workday. Regular movement is a requirement of our muscular system, as is exemplified in the process of sleep. We think of sleep as the time when we achieve maximum rest for our bodies and that movement is kept to a minimum. However, just the opposite is true. Have you ever awakened from a good night's sleep in exactly the same position as you fell asleep? Unlikely. The muscles of the body are often very active during sleep to release electrical energy and re-balance for the next day's use. So, in the workplace normal body movements should be made often, with comfort and efficiency.

The same goes for the visual system. The eyes are designed to focus at all distances and move in many directions. If there is a lack of movement – the eyes stay focused at one spot for an extended time – tension results and the muscles stiffen. If the eye movement is exaggerated for long periods of time, excess stress will result, which can in turn lead the body to attempt to relieve the stress. Normal eye movements can be achieved in short periods of time with alternate viewing tasks to reduce eye fatigue.

Office lighting

Office lighting is one of the most overlooked and under-emphasized components of the workplace. Lighting is something most of us don't pay much attention to – but our eyes are very much affected by it. Our work can be made much easier or harder by varying the kind of illumination in the work area. Our eyes are designed to work with a certain amount of light and to adjust to varying amounts of light. However, improper lighting can create a very stressful visual situation, leading to poor performance and worker discomfort.

There are four general types of lighting available in the workplace.

Ambient light is general, overall room lighting. It is usually in the form of direct recessed or ceiling-mounted fluorescent tubes. Ambient light is easiest on the eyes when it is indirect, with fixtures positioned to wash or bounce light off walls and ceilings. Too low ambient illumination can be gloomy. Balance is the key: avoid 'hot spots'.

Task lighting is used to individually light the worker's specific task area, such as lighting the copy for computer work. Ambient and task lighting are the two most essential kinds of lighting of the workplace. The trend is for reduced ambient light combined with adjustable task lighting. Most offices lack sufficient task lights. A good task light is the best combination of bright, flexible and directional.

Accent or directional lighting is used to light specific objects such as works of art, or to help balance room brightness levels. It is important to keep this understated and not competing with or distracting from ambient and task lights.

Natural light, or daylight, can come from a window, glass door, wall or skylight. Consider this a bonus – do not count on it. And also be cautious of the glare that this lighting can create on the display screen.

Here are some general considerations to address in the workplace. Many of the remedies of these problems will be addressed in Chapter 9.

Brightness/Contrast: The immediate work area – the area on which central vision is focused – should not be very different in brightness from the rest of the room. The brightness of the central field should never be less than that of the surrounding area; in some cases where continuous, intense work is necessary, it can be slightly greater. Too great a contrast between the central and peripheral visual areas is uncomfortable and interferes with vision and work.

Much of the consideration will depend on whether the VDT user is using a bright or dark background on their monitor. The dark background will dictate that the workplace lighting be subdued so as to not create a large difference in illumination between the screen and the office light. Conversely, if the background is light, a lighter office space is acceptable. However, it has been found that most offices are two to three times as bright as they should be for VDT use.

Glare: Glare is caused by a light source that is too bright or too close to the eyes or one that is positioned so that it reflects off surfaces in the work area. Older people are usually more sensitive to glare than are younger workers. Bare light bulbs should never be used in a work environment. The brightness and position of lights should be adjusted to avoid glare, and shiny finishes on furniture, walls, ceilings and other surfaces should be avoided. Pay attention to the colors selected for office furnishings and equipment because different colors reflect different amounts of light. For example, black reflects only 1% of visible light; dark blue, 8%; light blue, 55%; and light gray, 75%. Consult a lighting designer to help you pick out the best combinations of light and color for your office.

There are two main types of glare to consider: veiling glare and discomfort glare. Veiling glare comes from the reflections previously discussed, the most significant sources being the reflections from the glass of the VDT, as well as lightly colored clothing or work surfaces. Discomfort glare is caused by scattered light directly entering the eye. This often comes from outside light or peripheral overhead lighting sources. This source of glare can be determined by using a visor or other shield to block out the light from the direction of the light. If the eyes feel better, then the light source is a concern which should be addressed.

Color of light: Most lights are available in many colors, so many that it may be confusing. Different colors will create different moods ranging from very cool to very warm. The most widely-used fluorescent colors are cool white (a whiter light with a green/blue cast to it) and warm white (with a reddish or rosy cast). Although lighting color preferences can be very personal, warm white is recommended for general use. While cool white tubes are the most available, least expensive and therefore most used, they cast an unhealthy pallor over everything. When your skin looks sick, it is hard to feel good. They also tend to cost more to run over the long term.

Full spectrum lights come closest to nature's light by imitating the color rendition of the noonday sun. They can add a whole new sense of well being to the office environment. Full spectrum fluorescent lights give off a more uniform energy and display colors in a bright, realistic light. When purchasing them, ask for a 5,000 degree Kelvin lamp.

Lighting is something most of us don't pay much attention to – but our eyes are very much affected by it, and our work can be made much easier or harder by varying the kind of illumination in the work

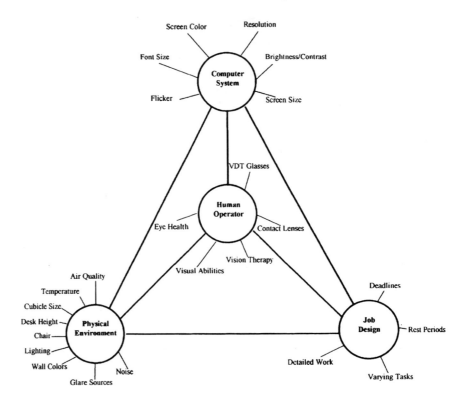

Figure 3.3 Factors affecting VDT stress (after Godnig, 1990).

area. Some of the technical aspects of lighting will be discussed in the next chapter. Recommendations on how to control lighting and the recommended levels for VDT work will be elaborated in Chapter 9.

References

Armstrong, T.J. (1991) Cumulative trauma disorders and office work. Unpublished document. Ann Arbor, MI: University of Michigan.

Bergqvist, U., Knave, B. and Wibom, R. (1994) Eye discomforts during work with visual display terminals. *Work With Display Units Book of Short Papers*, Vol. **1**, B13–B14.

Blustein, G., Rosenfeld, M. and Cluffreda, K. (1993) Does Dark Accommodation Really Change Following Sustained Near Fixation? Archives of the American Academy of Optometry, Dec.

Grandjean, E., Hunting, W. and Pidermann, M. (1983) VDT Workstation Design: Preferred Settings and Their Effects. *Human Factors*, **25**, 161–175.

Hill, S.G. and Kroemer, K.H.E. (1986) Preferred Declination of the Line of Sight. *Human Factors*, **28**, 2, 127–134.

Jaschinski-Kruza, W. (1988) Visual Strain during VDU Work: The effect of viewing and distance and dark focus. *Ergonomics*, **33**, 8, 1055–1063.

Kruegar, H. (1984) Visual Functions in Office – Including VDUs. *Ergonomics and Health in Modern Offices*. London: Taylor & Francis.

Lehmann, G. and Stier, F. (1961) Human and Equipment. In *Handbuch der Gesamten Arbeitsmedizn* (Vol. 1) p. 718–788. Berlin: Urban und Schwarzenberg.

Leibowitz, H.L. and Owens, D.A. (1975) Anomalous myopia and the intermediate dark focus of accommodation. *Science*, **189**, 646–648.

Murch, G. (1982) How Visible is your Display? *Electro-Opt Sys Design*, **14**, 42–49.

Owens, D.A. (1984) The resting state of the eyes. *American Scientist*, **72**, 378–387.

Rouse, M. *et al* (1994) Comparison of Accommodative Lag between Viewing Printed and Video Display Terminal Text. Archives of the American Academy of Optometry, Dec.

Sauter, S.L., Arndt, R. and Gottlieb, M. (1981) A Controlled Survey of Working Conditions and Health Problems of VDT Operators in the New York Times. Department of Preventive Medicine, University of Wisconsin, Madison.

Sheedy, J.E. (1994) How do eye problems rank with other VDU disorders? Presented at Work With Display Units, Fourth International Scientific Conference, University of Milan, Milan, Italy.

Chapter 4
Visual perception and VDTs

Introduction

There is much talk these days about the use of various types of input devices to relieve the stress of computer use. There is on-going work with voice activated software; a 'foot-mouse' that uses the feet to control the cursor; a 'ring' mouse which attaches to the finger and can be used in free space; and other unique variations which attempt to alter the repetitive motion which the computer user must endure. However, regardless of what type of input device is used, the output on computers will always involve the eyes and visual system. Therefore, it is imperative that we address the visual issues of viewing the VDT screen.

Just as we first had to learn a bit about the workings of the human visual system to better understand its make-up, we must now also address the electronic member of this tandem – the Visual Display Terminal itself. This is not meant to be an electronic treatise of the technical aspects of VDT technology. However, it is meant to present the basics of how VDT technology works so that you are able to purchase a monitor based on its technical abilities and your requirements and not the persuasive abilities of a salesman.

Of CRTs and LCDs

There are at present two general technologies available to create an image of the screen of a VDT. The original method is the Cathode Ray

Tube (CRT) and the newer is the Liquid Crystal Display (LCD). Both have unique characteristics which can be advantageous in certain viewing situations.

The CRT was developed in the early 1950s and uses a large vacuum tube with an internal electron gun to create an electron beam. The beam strikes the phosphor coating on the inside of the CRT face. It is then scanned, through the use of grid with electrical potentials, across the inside surface of the CRT. The image, therefore, is made up of a set of horizontal lines, each of which represents a thin slice of the image being drawn. The graphics controller in the computer must signal to the display after each line has been drawn; this is analogous to a 'carriage return' command to a printer. It must also signal to the display at the end of each full picture or frame, this being analogous to a 'page feed' command to the printer. Unlike the text on a printed page, which is permanent, each frame on a CRT display is no more than a brief flash of lighted phosphor dots (also referred to as picture elements, or 'pixels') which disappear in a fraction of a second, depending on the persistence of the screen phosphor. The image does not disappear because the controller and the display together redraw, or refresh each frame continuously, many times each second.

If, for example, the screen is capable of being scanned 72 times a second (refresh rate of 72 Hertz (Hz)), the electron gun fires over 37 million times a second. This assumes a screen which has a pixel arrangement of 832 pixels horizontally by 624 vertically (832 × 624). This resulting figure is called the bandwidth for the VDT, in this case being 37 megahertz (832 × 624 × 72). In principal, the greater the bandwidth, the better the monitor. However, for some monitors, the picture tube may be the limiting factor affecting display performance, in which case there is little point in increasing the bandwidth beyond a certain point.

A color CRT display has an additional level of complexity. In order to achieve color, the inside of the CRT face contains many red, green and blue phosphor dots. The phosphor dots are each activated by a separate electron beam. As the three electron beams scan across the display, a 'shadow mask' is used to separate the three beams and allow each to stimulate their respective phosphor dots. The shadow mask is a curved metal plate with numerous small round holes through which the electron beams stimulate the appropriate phosphor dots.

In a color monitor, the spacing between the holes in the shadow mask is called the 'dot pitch'. The size of the dot pitch is an important characteristic in describing the ability of a color monitor to display detail; smaller dot pitches are capable of displaying more detail. Also, since a portion of the electron beam is absorbed by the shadow mask instead of the phosphor, color displays often are not as capable of attaining the same luminance as monochrome monitors. Figure 4.1 shows how pixels are arranged and dot pitch is measured.

The make-up of the LCD panel is very much different from the CRT. A thin film of liquid crystal is placed between two pieces of glass or transparent plastic. These plates are usually manufactured with transparent electrodes, typically made of indium tin oxide, that make it possible to apply an electric field across small areas of the film of liquid crystal. Polarizing filters are usually placed on one or both

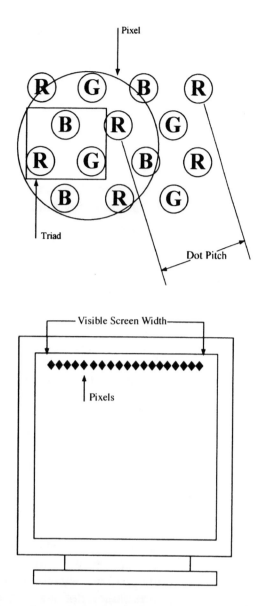

Figure 4.1 The relationship between dot pitch, triad and pixel.

sides of the glass to polarize the light entering and leaving the crystal. Usually these polarizers are crossed which means that, normally, no light would be able to pass through the display. The liquid crystal, however, will modify the polarization of the light in some way that is dependent on the electric field being applied to it. Therefore, it is possible to dynamically create spots where light will get through and spots where it will not.

What is liquid crystal? There are three common states of matter that most people know about: solid, liquid, and gas. Liquid crystal is a fourth 'state' that certain kinds of matter can enter into under the right conditions. The molecules in solids exhibit both positional and orientational order – in other words, the molecules are constrained to point only in certain directions and to be only in certain positions with respect to each other. In liquids, the molecules do not have any positional or orientational order – the direction the molecules point and their positions are random. The liquid crystal 'phase' exists between the solid and the liquid phase; the molecules in liquid crystal do not exhibit any positional order, but they do possess a certain degree of orientational order. The molecules do not all point the same direction all the time. They merely tend to point more in one direction over time than other directions.

Today's LCDs come mostly in two 'flavors': passive and active. The less expensive passive matrix displays trade off picture quality, view angle, and response time with power requirements and manufacturing cost. Active matrix displays have superior picture quality and viewing characteristics, but need more power to run and are much more expensive to fabricate. Liquid crystal displays show great potential for future growth and improvement.

LCDs offer several advantages over traditional cathode-ray tube displays that make them ideal for several applications. LCDs are flat, and they use only a fraction of the power required by CRTs. They are easier to read and more pleasant to work with for long periods of time than most ordinary video monitors. There are several tradeoffs as well, such as limited view angle, brightness, and contrast, not to mention high manufacturing cost. As research continues, these limitations are slowly becoming less significant. One study (Miyao *et al*., 1993) examined which type of display is better. Middle-aged subjects showed slower reading speed than young subjects. Then, accommodative function showed that non-backlit LCDs reduced focusing speed among young subjects and reading performance in middle-aged subjects. Since middle-aged workers have more difficulties than young workers, they should have their vision properly corrected for VDT viewing, more appropriate displays and a more comfortable illuminance environment than young workers.

Display screen variables

There are significant differences in viewing a VDT screen and viewing a piece of paper. Resolution, legibility, brightness, contrast, color, and flicker and jitter are the primary factors concerning the presentation of the image. Let's examine each of these individually and see how they affect the image produced.

Resolution defines how finely the image is reproduced by the triad array and is specified either as the horizontal and vertical size of the image in pixels (e.g. 1024 × 768), or as the number of horizontal pixels (or dots) per inch (e.g. 72 dpi). Often the dpi is not specified by the manufacturer. However, it can be easily calculated horizontally and vertically on the screen by dividing the reported number of pixels by the screen dimension in inches.

It is important to understand that there are two kinds of resolution: addressable resolution and image resolution. Addressable resolution refers to the signal output by the graphics controller where each pixel has a unique address in the controller's video memory. The addressable resolution defines the maximum image detail available. Image resolution describes how sharp and crisp the image will appear on the screen. This depends to some extent on the number of triads per pixel. Dot pitches for computer displays are typically in the range of 0.25mm to 0.31mm and the total number of triads should be equal to or, ideally, greater than the addressable resolution to ensure that characters are well-defined and well formed.

The dot pitch required to support a particular addressable resolution can be estimated by dividing the visible screen width by the horizontal addressable resolution. For example, say we require a 17-inch monitor to display 1024 × 768 resolution, the required minimum dot pitch is estimated as follows: 315mm (visible display width) divided by 1024 (horizontal resolution) = 0.31mm. Inadequate dot pitch results not only in degraded resolution but can also produce disturbing wispy and wavy bands superimposed on the display. As a rule of thumb, the finer the pitch, the better (and more expensive), all other things being equal.

However, caution should be exercised when evaluating monitors by manufacturer resolution numbers alone. The dpi can be used as a meaningful comparison of monitors as long as the character sizes are the same. If, however, a denser pixel arrangement only results in smaller letters on the screen because the pixels allocated to each letter are the same with either pixel density, then there are no gains in image resolution. In general, and in most applications, there are resolution advantages to using more pixels. Today, a standard dot pitch of 0.31mm is considered acceptable, 0.28mm is preferable and 0.25mm is exceptional.

Using a monitor with better resolution does more than just make the display images more pleasing – it can improve performance and reduce discomfort. In one study (Sheedy, 1992) reading performance and visual symptoms were measured on a group of subjects who read stories on a monitor with 116 × 118 dpi and on a VGA monitor with 73 × 80 dpi. The higher resolution monitor resulted in a 17% faster reading rate and only 38% as many symptoms. It seems clear that greater pixel densities result in better performance and comfort.

Legibility is a term of measure which is broader than resolution. It represents the ability to be deciphered or read. The ANSI/HFS 1996 standard defines legibility as 'the ability for unambiguous identification of single characters that may be presented in a non-contextual format'.

The intended viewing distance must be known before any calculations may be made to determine if the size of a character is suitable. This angular height is expressed in minutes of arc (a minute is 1/60th of a degree). Preferably, if the task requires a significant amount of reading for context, the workstation should permit the display to be used at a distance where the character height is approximately 20–22 minutes of arc. The minimum recommended size is 16 minutes of arc. For application where reading is incidental to the task, smaller characters may be used.

Legibility is becoming more critical as higher resolutions, e.g. 1024 × 768 and 1280 × 1024, are coming into widespread use. For example, a 21-inch monitor can deliver the minimum character size at 1280 × 1024 resolution. A 17-inch monitor delivers just over the minimum at 1024 × 768 resolution, whereas the character size on the 15-inch monitor is on the minimum at that same resolution. A similar analysis can be done for a particular user's requirements taking into account the actual viewing distances and character sets.

The measurements in Table 4.1 are based on a Times Roman 12-point font and a constant viewing distance of 60cm (24in). As a rule of thumb, 21- or 20-inch monitors should be used for 1280 × 1024 resolution, 17-inch monitors for 1024 × 768, 15- and 14-inch monitors for 800 × 600. A relatively simple test can be applied to determine whether the character size/resolution is appropriate to the task and the person. The user should back away from their computer screen

Table 4.1 *The size variation for the character 'E' based on resolution and character height for various size displays*

Display size (in)	Resolution	Character height (mm)	Character size (min. of arc)
15	1024 × 768	2.81	16.08
17	1024 × 768	3.16	18.11
21	1024 × 768	4.06	23.26
21	1280 × 1024	3.08	17.65

and determine the farthest distance at which they can just barely identify the (preferred) text on the screen. This establishes the threshold size for that person. One-third of that distance represents the distance at which the text is three times the threshold for that person's visual comfort – and also the maximum distance at which the person should be viewing the screen. For example, if the maximum distance is 6ft then 24in (1/3 of 6) is the recommended comfortable viewing distance. This evaluation method assumes that no 'reading' glasses are used for that distance.

The **brightness/contrast** concerns of the VDT user are related to the surrounding illumination of the workspace. Brightness is a combination of luminance and chrominance. The maximum luminance ratio of the task to the immediate surroundings should not exceed 3:1. Any sources of glare on the surface of the screen will decrease the contrast proportionally. You might want to refer to the Glossary to clarify some terms which are used in the discussions of lighting.

The requirements of ISO9241-3 are that the display should be capable of a luminance of at least 35 cd/m^2 (candelas per square meter) and that the luminance of object or areas which are frequently viewed in sequence (i.e. screen, document, etc.) should not differ by more than a ratio of 10:1. Higher luminance ratios between the frequently viewed areas or objects and the general surroundings (e.g. desk area and room walls) are acceptable. A luminance level of 35 cd/m^2 would be considered fairly dim and 100 cd/m^2 would be near the maximum brightness level on a modern VDT. The eye does not react very quickly to changes in light levels. It can accommodate a wide range of illumination levels but it does take time to adjust to a particular level. If the eye is forced to adjust frequently between areas with large differences in illumination levels then eye fatigue and headache may result.

Visual acuity is measured by determining the minimum contrast between light and dark lines. There is no significant improvement in acuity above around 30 cd/m^2; contrast becomes the important factor after that point. Contrast is limited by the maximum practical difference in luminance between the light and dark areas of the display. The maximum image brightness is limited by defocusing and 'blooming' effects which begin to make the image increasingly fuzzy as brightness is increased. In addition, the darkest possible minimum brightness (i.e. screen is turned off) is limited by ambient glare and reflection and the display tube technology.

In practice the user should not set the brightness too high and should attempt to match the brightness of the display to the brightness of the immediate work area. The best image will result with an adequate text size (16 minutes of arc or greater) and a brightness level which does not produce defocusing effects. The display brightness and contrast might need to be adjusted during the day to match ambient lighting conditions should they change.

There are a few commonly used configurations for VDT screen presentations: dark background with light letters, including green and amber (normal polarity); light background with dark letters (reverse polarity); colored background with light or dark letters. Each of these configurations has a specific contrast requirement based on the type of monitor used and the environment of the workstation. The older, dark background with light letters has been used in office situations over a long period of time. As we discussed in the previous chapter, the lighting conditions of the office environment created an unnatural situation for normal and comfortable viewing. Reflections off the screen can decrease the contrast of the letters to the background and make the letters 'wash out'.

More recently, a reverse polarity was available due to the faster refresh rates of the monitors. This larger, brighter background calls for a re-evaluation of the office lighting factors which have been previously discussed. Slightly lighter offices can achieve acceptable ranges if the screen is bright. Also, the glare from various sources may not become as significant as it had been with the darker backgrounds. Caution should, however, be exercised in viewing these lighter backgrounds. If they are in view of the peripheral retina, such as in an adjacent office area, flicker can be perceived much more easily.

Display screen users often ask which color is the 'best' to view on the screen. The early issues concerning color were in terms of monochrome monitors which had the choice of amber, green or white characters. The green was the original text color and was the most difficult color to resolve, which is why amber became the more popular choice. There can be arguments made for most display screen colors which support their use. For example, the eye does not have to focus as much for blue characters; red or amber characters are clearer because the red wavelengths are not as subject to scatter as are the blue wavelengths; green characters balance the positive and negative features of blue and red characters and will allow the eye to focus the same as if it were a white stimulus.

Some older studies have shown that colors do affect the way we read, whether it be VDT or paper reading. Patterson and Tinker (1931) found that there was a 16% increase in visual efficiency and a 14.7% increase in legibility in printed material with black letters on white background when compared to red on green, green on white, blue on white and blue on yellow. They also found that eye movements were more efficient with black on white when compared to these same color combinations. However, in the final analysis, the relative advantages of one color over another are small and tend to cancel one another. When it comes to selecting a monochrome monitor, the color of the monitor is relatively unimportant compared to the other aspects such as contrast polarity and resolution.

The full colored display screens have been a source of much curiosity and should not pose a significant problem, as long as proper contrast can be maintained between the letters and the background. For example, a black letter on a pale blue background is probably acceptable where a red letter against a yellow background will not offer enough contrast to see comfortably. This is because our visual systems process color and luminance differently. The visual system can detect fine variations in luminance but only coarse variations in color. Detecting red text on a yellow background is difficult because there is little difference between the luminance of the text and that of the background.

Recall that the pixels required to make up the image for a color screen are composed of red, green and blue. This means that there is a loss of resolution with the color presentation because the energy to generate the image must be split into thirds. Therefore, it is essential that the contrast and brightness considerations be addressed before discussion of which color might be optimum for display screens. However, the technology of creating a comfortable viewing image is creating displays which are 'easy' on the eyes.

The discussion of **flicker** has been a concern of VDT users and manufacturers, as well as researchers, for many years. Human sensitivity to flicker is usually expressed as the critical fusion frequency (CFF). This is the flicker frequency rate beyond which we can no longer perceive the flicker (or refresh, in the case of VDTs). There is an individual variation in the CFF and it also depends upon various aspects of the visual stimulus. For most common viewing conditions, the CFF is in the range of 30–50 Hz. Since this is below the computer screen refresh rate, we do not usually perceive the flicker. However, there are some conditions in which flicker can be more easily perceived. For example, screen brightness can make a difference because a brighter stimulus results in a higher CFF. It is common for people to perceive the screen flicker on a white background if the brightness is adjusted high. This problem can be easily eliminated by adjusting the screen brightness lower.

Also, screen location (retinal image position) is a factor. The peripheral retina has a higher CFF and is therefore more sensitive to flicker than the central retina. Most of us can perceive the flicker of a white background screen by looking to the side of it. The worker who continuously views a reference document placed to the side of the screen is more at risk for seeing flicker. Rearrangement of the work station and/or brightness control can help to alleviate this problem.

As previously mentioned, the larger the source of the light, the higher the CFF. Therefore, we are more likely to perceive the flicker on a white background screen than a dark background screen. Although this single factor favors the use of dark background screens, the balance of other factors strongly favors the use of white background screens. Studies have shown that even flicker that is beyond our ability to perceive may create a potential problem. Berman *et al.* (1991) recorded flicker responses from the retina at frequencies as high as the 125–160 range. Another study by Harwood and Foley (1987) showed that the flicker, whether perceived or not, could cause fatigue of the visual system, as well as an increase in symptoms.

A similar but different phenomenon is **jitter** of the image. Jitter is the perception of small unintended variations in the position of the lines which make up the picture. ISO 9241-3 suggests that the image should appear 'stable' and defines this to be a geometric movement of picture elements of no more than 0.0002mm per mm of design viewing distance. For a viewing distance of 600mm this would mean a movement of no more than 0.12mm.

In practice jitter can be very apparent and annoying. It is particularly apparent where displays are driven by an interlaced signal. Interlacing is a technique which is designed to reduce apparent screen flicker by doubling field refresh without the expense of faster electronics. This is done by drawing first all the odd numbered lines and then filling in the even lines so that a complete picture is composed of two interlaced fields to make up a full picture. This makes it possible to deliver an interlaced 86Hz field refresh from a display which is only capable of 43Hz frame (non-interlace) refresh. In some applications, such as television, this works fine. With text-based computer applications and especially graphical user interfaces, this works less well and simply swaps one bad effect (flicker) for another (jitter). Computer images tend to be very precise and the time delay between drawing odd and even lines introduced by interlacing tends to accentuate small movements creating a 'sparkling' or 'vibrating' effect over the whole image.

Purchasing a monitor

All of the preceding information is designed to inform you of some of the technical aspects of the display screen and monitor. However, the ultimate test comes when you must purchase the best monitor for the particular workplace environment. Reading advertisements can be confusing and technical terms are bandied about with the purpose of creating confusion. Let's take a look at some of the factors you should consider in the purchase of the monitor and what you get for your money. There are some questions you should be asking regarding the various monitors. These include:

- Color quality: How accurately does the monitor reproduce colors?
- Image quality: Is the image well focused? Do the three colors (red, green and blue) that make up all images remain aligned with each other over the whole screen? Is the image free from geometric and other distortions?
- Ergonomics: Are the controls well placed and easy to use? Do the controls feel sound? Is the monitor easy to tilt and swivel? How much space is required for this movement? How much space does the base occupy?
- Controls: How many screen settings are adjustable? Are the on-screen display and controls easy to understand? Is the on-screen display animated and/or does it provide feedback?
- Power conservation: How conservative is the unit's power consumption? Is there a deep-sleep mode?

Next you will want to decide what size of monitor to get. Your choices are generally 14, 15, 17, or 21-inch. Consider the kinds of applications you use, the amount of money you're willing (or able) to spend, and available desk space.

Fourteen-inch monitors were the standard a few years ago but the technology has advanced (and the price has declined) so that your minimum purchase should probably be a 15-inch. If you spend most of your time working at resolutions no higher than 800 × 600, a 15-inch monitor should provide you with enough room for applications such as word processing and database entry. You're probably better off with a 17-inch monitor if you work at resolutions of 1024 × 768 and higher. A monitor of this size also offers advantages at lower resolutions – the extra screen real estate gives you more room to work with multiple windows, applications minimized to icons and large spreadsheets. If you use desktop publishing, graphics or CAD applications, a top-of-the-line 17-inch model offers a viable alternative to a 21-inch model. Also, working on a 21-inch monitor at 1280 × 1024 will require a minimum of 75Hz.

Bigger displays come in bigger boxes, so make sure you've got the desk space to accommodate the size you choose. Another point to remember: while a monitor may be labelled a 15- or 17-inch unit, the actual image size may be an inch or two smaller. Be sure to make note of the 'viewable screen area' measurement.

Make sure the monitor you buy supports the same refresh rates as your video card. Also make sure that the refresh rates supported by the monitor are high enough, since a low refresh rate can cause eyestrain. The recommended refresh rate is 72Hz or higher; less than 70Hz will result in obvious flicker. If your video card supports DPMS (Device Power Management Standard), look for a DPMS-compliant monitor. When paired, the two will power down after a period of inactivity. Most monitors that meet the EPA's Energy Star guidelines are DPMS-compliant.

If possible, don't buy a monitor sight unseen. Find a store where you can test and compare a number of models. Check for margins of black around the screen edges. Ideally, the image should fill the screen from top to bottom and left to right, and should be adjustable. Also keep an eye out for bowed and pinched edges – see if you can fix the problem by using the pincushion or barrel controls. Check for color distortion and poor convergence on the edges of the screen, and see if controls improve what you see.

You might try this little test: fill the entire screen with many letter 'E's. Check to see that the image covers the entire area, including the corners. Make sure that the letters in the corners are as focused and bright as the ones in the center. Also, put up a white background on each monitor and adjust the brightness and contrast. This will help you judge each screen's overall brightness. Pick one or two basic images and put the same picture up on each monitor. Compare the color contrast, the brightness intensity, and the picture crispness and quality. Finally, put up white text on a dark screen from the C: prompt. Look at the text closely to make sure that the convergence (when red, blue and green rays come together to create white) is good and that no colors are bleeding out at the characters' edges.

Below is a list of specifications which often appear in the technical section of a monitor advertisement. We'll compare three different monitors and see what numbers actually matter.

In fact, all three of these monitors are well suited for almost any workplace application. There are some minor differences between them which may sway you in one direction or another but, for the

Table 4.2 A comparison of various attributes of available monitors

Specification	Monitor A	Monitor B	Monitor C
CRT type	15in non-interlaced	15in non-interlaced	15in non-interlaced
Viewable Screen Area	13.8in	14.0in	13.8in
Refresh Rates (recommended)	1024 × 768 60–72Hz	1024 × 768 75Hz	1024 × 768 75Hz
Scanning Frequency	H: 31–65Hz V: 55–160Hz	H: 30–61kHz V: 50–90Hz	H: 30–64kHz V: 50–90Hz
Video Bandwidth	85MHz	75MHz	100MHz
Display Colors	unlimited	unlimited	16 million
Dot Pitch	0.28mm	0.27mm	0.28mm
Screen Surface	Anti-static, non-glare	Anti-static, silica-coat	Anti-static, non-glare
Controls	front panel; on-screen	front panel	front panel

most part, they are insignificant. For example, monitor B has the better dot pitch and larger viewable screen area but monitor C has the better bandwidth. Monitor A has on-screen controls which might be handy if various users have different visual needs and adjustments are made frequently. There is no practical difference in having 16 million colors versus unlimited since the visual system can detect about 7 million shades of color.

Some terms which you should be familiar with when making judgments on a monitor appear in the Glossary at the end of this book.

In addition to just purchasing a monitor, the image quality you perceive is affected by the other major component of image production – the graphics accelerator. This is the card that is responsible for controlling the 'heart' of the video system. The card you choose will probably use either VRAM (Video Random Access Memory), which is more expensive but fast, or DRAM which is moderately priced and slower. DRAM cards typically come with 1MB or 2MB of memory. VRAM cards, used most often for CAD and desktop publishing, come in 2MB, 4MB or even 8MB configurations. A few companies are using a new memory type, EDO (Extended Data Out) RAM, which costs less than VRAM but provides better performance than DRAM.

Make sure you select a card that supports the same bus as your computer. Also consider how and when you'll want to upgrade your computer. Older machines often use an ISA bus, while many 486 PCs use VL-Bus. However, the future standard is PCI, which is available now on most Pentium systems. A card with a 64-bit controller that uses 2MB of DRAM generally costs less than $200, while a VRAM version with 2MB of memory costs around $300. Some high-end VRAM cards with 8MB of memory cost $500 or more (1995 prices).

High refresh rates eliminate screen flicker. For most users a rate of 72Hz to 75Hz is enough to achieve the desired results. Some cards support refresh rates of up to 120Hz. If you need this kind of rate to provide an extremely clear and stable image, make sure that your monitor can support it. Before you rush to the store to buy a graphics accelerator card, look carefully at your current system and how you use it. Consider your color depth and resolution requirements. If you have a 14- or 15-inch monitor, you will probably use 800 × 600 resolution; the preferred resolution for a 17-inch monitor is 1024 × 768. Power users with a 21-inch display will want 1280 × 1024. The higher the resolution you want, the more memory you'll need. If you choose a card with only 1MB of RAM, your 64-bit controller will actually run as a 32-bit device, since a megabyte of RAM is only 32 bits wide. If you later decide you need more memory, many cards can be easily upgraded. Many cards also include utilities that let you change color depth and resolution on the fly.

There are some general rules of thumb to keep in mind when looking at a display screen.

1. Screen resolution: this is, especially important where extended viewing times are expected. VGA displays should have a resolution of 75 dpi or more. Compare this number to most printers which have dpi starting at the 300 range and go higher from there.
2. Black characters on a white background is probably the best. However, other combinations can be comfortable as long as the brightness contrast between the characters and the background is high. It is best to avoid dark backgrounds.
3. The size of the text should be three times the size of the smallest text that can be discerned. See the '3X' rule above.
4. Monochrome displays usually have better resolution than color. If the job does not require color, it is often best to use a monochrome monitor (no fun at all!).
5. The higher the refresh rate, the better. At least 70Hz is the minimum to consider.
6. For color monitors, smaller dot pitches (0.28mm or less) are desirable.
7. Adjust the screen contrast so that character definition and resolution is maximized.
8. The screen brightness should be adjusted to match the general background brightness of the room. This is much easier to do with light background screens.
9. Gray scale is desirable to obtain better character resolution and smoother-appearing borders of the characters.
10. If there are problems with after-images, turn down the screen brightness, use dark characters on a light background and avoid green characters on a black background.
11. It is generally desirable to have a monitor with capability of displaying high screen luminance. This ensures being able to match the luminance of the screen to that of the visual surroundings. Also, if an anti-reflection filter is going to be used to enhance contrast, a higher maximum screen luminance enables compensation for the luminance decrease caused by the filter.

There are many options and offerings which can be recommended for the VDT users. The only way to judge a monitor is by looking at it and working with it using your typical applications. Ultimately, however, the determination of whether a monitor is good or not rests in the eyes of the beholder.

References

Berman, S.M., Greenhouse, D.S., Bailey, J.L., Clear, R.D. and Raasch, T.W. (1991) Electroretinogram responses to video displays, fluorescent lighting, and other high frequency sources. *Optometry and Visual Science* **68 (8)**, 645–662.

Harwood, K. and Foley, P. (1987) Temporal Resolution: An Insight into the Video Display Terminal 'problem'. *Human Factors* **29 (4)**; 447–452.

Miyao, M., Ishihara, S., Furuta, M., Kondo, T., Sakakibara, H., Kashiwamata, M. and Yamada, S. (1993) Comparison of readability between liquid crystal displays and cathode-ray tubes. *Nippon Eiseigaku Zasshi*, **48:3**, 746–51, Aug.

Patterson, D.G. and Tinker, M.A. (1931) Black versus white print. *Journal of Applied Psychology*; **15**, 248–251.

Sheedy, J.E. (1992) Reading Performance and Visual Comfort on a High Resolution Monitor Compared to a VGA Monitor. *Journal of Electronic Imaging* **1(4)**, 405–410.

Chapter 5

Computer Vision Syndrome

Introduction

Because VDT use is such a highly visually-demanding task, vision problems and symptoms are very common. Most studies indicate that VDT operators report more eye-related problems than non-VDT office workers. A number of investigators (Smith *et al.*, 1981; Yamamoto, 1987; Dain *et al.*, 1988; Collins *et al.*, 1991) have indicated that visual symptoms occur in 75–90% of VDT workers. A study released by NIOSH showed that only 22% of VDT workers have musculoskeletal disorders.

A survey of optometrists (Sheedy, 1992) indicated that 10 million primary care eye examinations are given annually in the United States primarily because of visual problems at VDTs. This study eventually culminated in the compilation of the series of symptoms which are now collectively known as Computer Vision Syndrome (CVS). This condition most often occurs when the viewing demand of the task exceeds the visual abilities of the VDT user. The American Optometric Association defines CVS as that 'complex of eye and vision problems related to near work which are experienced during or related to computer use'. The symptoms can vary but mostly include eyestrain, headaches, blurred vision (distance and/or near), dry and irritated eyes, slow refocusing, neck and/or backache, light sensitivity, double vision and color distortion.

The causes for the inefficiencies and the visual symptoms are a combination of individual visual problems and poor office ergonomics. Poor office ergonomics can be further divided into poor workplace conditions and improper work habits. The above mentioned survey also concluded that two-thirds of the complaints were related to vision problems while one-third was due to environmental factors. Many people have marginal vision disorders which do not cause symptoms when performing less demanding visual tasks. However, it has also been shown that VDT users also have a higher incidence of complaints than non-VDT users in the same environment (Udo et al., 1991).

Let's review these symptoms and see if we can determine how they arise and how they may be addressed, both visually and environmentally.

Eyestrain

The eye care professions have no set definition of 'eyestrain'. We have definitions for numerous conditions which occur throughout the visual system but eyestrain is not one of them. One tends to think of a muscle strain as what would happen to a muscle if it were overworked. In fact, it is rarely a tired muscle which causes many of these complaints.

The term used professionally for 'eyestrain' is *asthenopia* (AS-then-OH-pee-ah), which itself is a rather vague term. The visual science dictionary defines asthenopia as the subjective complaint of uncomfortable, painful and irritable vision. It then gives twenty-four different types of asthenopia based on various causes. Because of its subjectivity, however, it can have a myriad of meanings to any number of people. Asthenopia can be caused from such problems as focusing spasm, different vision in each eye, astigmatism, hyperopia, myopia, excess light, voluntary focusing, eye coordination difficulties, and more.

In the VDT environment, eyestrain in all of its manifestations can also be caused by a number of different environmental conditions. It is unfortunate that this is the most common complaint of VDT workers because it is also the most difficult to evaluate. It is very difficult for an eye examination in an eye-care professional's office to factor in a VDT worker's environmental conditions. Many people may consider the tiring of their eyes as the 'eyestrain' condition. This is most often caused by a condition known as convergence insufficiency and can be easily treated with a simple vision therapy program.

When confronted with the complaint of 'eyestrain', it would be prudent to have further testing done to determine the exact cause of the complaint.

Headaches

Headaches are another of those asthenopic symptoms and are the primary reason why most people seek an eye examination. They are also one of the most difficult maladies to diagnose and treat effectively. Headaches are reported at least once a month by 76% of women and 57% of men. There are numerous types of headaches and they can be caused by a number of different conditions. The International Headache Society classifies headaches in the following categories:

- Migraine
- Tension-type
- Non-vascular intracranial disorder
- Substance withdrawal

- Cluster
- Misc. unassociated w/structural lesion
- Head trauma
- Vascular disorders
- Noncephalic infection
- Metabolic disorder
- Facial pain
- Cranial neuralgia

It is beyond the scope of this book to delve into the various headache conditions and their origins. However, it would serve our purpose to distinguish between visual and non-visual origin headaches and what might be the source of the symptom.

Visual headaches most often occur toward the front of the head (there are a few exceptions to this); occur most often toward the middle or end of the day; do not appear upon awakening; do not produce visual 'auras' of flashing lights; often occur in a different pattern (or not at all) on weekends than during the week; can occur on one side of the head more than the other and other more general symptoms. It is, therefore, imperative to elicit a thorough case history to distinguish the type of headache involved. The worker should be queried about the time of onset, location of the pain, frequency, duration, severity and precipitating factors such as stress, certain foods or medications. Associated signs and symptoms such as nausea, vomiting, light sensitivity and noise sensitivity should also be noted.

Often a worker will complain of a 'migraine' headache. However, migraines are a very specific type of headache and have an organic, not visual, cause. There is no clinical diagnostic test to establish the presence of a migraine headache so extensive laboratory tests would be appropriate. The worker should be referred for a neurological evaluation after all other variables have been accounted for.

VDT workers probably get tension-type headaches. These can be precipitated by many forms of stress, including anxiety and depression; numerous eye conditions, including astigmatism and hyperopia; improper workplace conditions, including glare, poor lighting, and improper workstation setup. These types of headaches are mild to moderate in intensity, often occur on either or both sides of the head, are not aggravated by physical activity, develop during the early to mid part of the day, last from 30 minutes to the rest of the day, and are relieved by rest or sleep. Chronic tension headaches vary somewhat from this; they have the same overall symptoms but occur much more frequently.

Visual and environmental conditions are the first places to look for a solution to a headache problem. If all obvious factors have been considered, medical management is in order, often starting with a complete eye examination to rule out the visual cause.

Blurred vision

Visual acuity is the ability to distinguish between two distinctive points at a particular distance. This requires the image formed on the retina to be well circumscribed and distinct. If the image focuses in front of or behind the retina, it will strike the retina in an unfocused state, creating the subjective symptom of blur. This process is true for all distances with the viewing range of the human eye, which we routinely consider from within 20 feet to 16 inches.

We consider the 20-foot distance optical 'infinity' due to the angulation of the light rays which emanate from that point. Whenever we direct our gaze to some point within 20 feet, we must activate our focusing mechanism to increase the focal power of the eye and regain the clear image on the retina. The ability of the eye to change its focal power is called accommodation and is dependent upon age. Therefore, we must consider many factors when discussing the accommodative ability of the individual.

Blurred vision symptoms can result from refractive error (e.g. hyperopia, myopia, astigmatism), improper prescription lenses, presbyopia or other focusing disorders. Wiggins and Daum (1991) found that small amounts of refractive error contributed to the visual discomfort of VDT users. Considering the working environment, blurred images can also arise from a dirty screen, poor viewing angle, reflected glare or a poor quality or defective monitor. All of these factors should be considered when this symptom occurs.

Recall our discussion of the lag of accommodation in a previous chapter. While viewing an object at a near (or intermediate) viewing distance of less than 20 feet, the eyes must accommodate. The point of focus is often not directly at the point of the object but usually behind it at some distance. As the VDT worker views the task for an extended period of time, the lag of accommodation increases, often leading to a subjective symptom of blur. The eyes must then expend more effort to pull the focusing point back to the screen. If this is accomplished with enough effort, then the symptom might become a headache; if not accomplished well enough, blurred vision might be the symptom.

A condition known as 'transient' myopia has been shown to be more prevalent in a population of computer users. One study (Luberto *et al.*, 1989) found that 20% of VDT workers had a nearsighted tendency toward the end of their work shift. This was also confirmed by Watten and Lie (1992) who found 30 VDT workers had this myopic trend after two to four hours of work. However, another study of transient myopia (Rosenfeld and Ciuffreda, 1994) showed that this condition also occurs after normal near-point viewing of a printed target. Studies showing permanent myopic changes have not shown this to be a concern at this time. However, many of those studies suffer from a lack of adequate control groups and low numbers of population tested.

Glare is also a concern because of the eye attending to the glare image rather than the screen image. If a specular reflection is noticeable on the screen, the eye will attempt to focus on it. The image of the glare source will appear to be somewhere behind the screen (much like your image is reflected in a mirror) and the screen image can appear blurred. This can become more noticeable as the usage time is increased.

Dry and irritated eyes

The front surface of the eye is covered with a tissue which consists of many glands. These glands secrete the tears that cover the eye surface and keep the eye moist, which is necessary for normal eye function. The tears help to maintain the proper oxygen balance of the external eye structures and to keep the optical properties of the eye

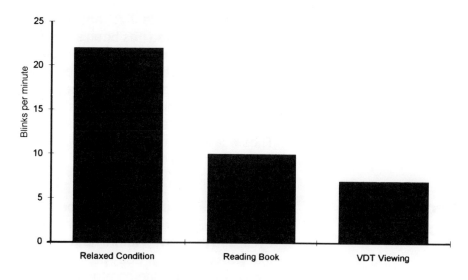

Figure 5.1 The variation of blink eye rate on various viewing tasks (Tsubota and Nakamore, 1993).

sharp. The normal tear layer is cleaned off and refreshed by the blinking action of the eyelids.

The blink reflex is one of the fastest reflexes in the body and is present at birth. However, our blink rate varies with different activities – faster when we are very active, slower when we are sedate or concentrating. Yaginuma *et al.* (1990) measured the blink rate and tearing on four VDT workers and noted that the blink rate dropped very significantly during work at a VDT compared to before and after work. There was no significant change in tearing. Patel *et al.* (1991) measured blink rate by directly observing a group of 16 subjects. The mean blink rate during conversation was 18.4 blinks per minute, and during VDT use it dropped to 3.6 – more than a five-fold decrease! Tsubota and Nakamori (1993) measured blink rates on 104 office workers. The mean blink rates were 22 blinks per minute under relaxed conditions, 10 while reading a book on a table, and seven while viewing text on a VDT. Their data support the fact that blink rate decreases during VDT use, but also show that other tasks can decrease the blink rate.

Possible explanations for the decreased blink rate include concentration on the task or a relatively limited range of eye movements. Although both book reading and VDT work result in significantly decreased blink rates, a difference between them is that VDT work usually requires a higher gaze angle, resulting in an increased rate of tear evaporation. Tsubota and Nakamori (1993) measured a mean exposed eye surface of $2.2cm^2$ while subjects were relaxed, $1.2cm^2$ while reading a book on the table, and $2.3cm^2$ while working at a VDT. The size of the eye opening is related to the direction of gaze – as we gaze higher, the eyes open wider. Since the primary route of tear elimination is through evaporation and the amount of evaporation roughly relates to eye opening, the higher gaze angle when viewing a VDT screen results in faster tear loss. It is also likely that the higher gaze angle results in a greater percentage of blinks that are incomplete. It has been suggested that incomplete blinks are not effective because the tear layer being replenished is 'defective' and not a full tear layer. The exposed ocular surface area has been shown to be one of the most important indices of visual ergonomics (Sotoyama et al., 1995).

Office air environment is often low in humidity and can contain contaminants. This has been noted as the cause of 'Sick Building Syndrome'. Additionally, the static electricity generated by the display screen itself attracts dust particles into the immediate area. These can also contribute to particulate matter entering the eyes, leading to dry eye symptoms.

Neck and/or backache

This book is about the visual aspects of VDT use, so why a section on neck and back problems? It is often heard in medical circles that 'the eyes lead the body'. Nature has made our visual system so dominant that we will alter our body posture to accommodate any deficiency in the way we see. One common example of a postural response to a visual stress is the act of squinting. When light enters the eye, it may be out of focus. The eye detects this blur and the person will squint their eyes to increase the depth of focus, thereby creating a sharper image on the retina. This demonstrates how the body will change to make the vision clearer. Galinsky *et al.* (1993) found that subjects monitoring a visual display reported greater subjective fatigue than those monitoring an auditory display.

A similar situation can be seen in many office situations where the vision of a worker is compromised and they must adapt their posture to ease the strain on the visual system. If an older worker is using glasses (single vision) which are designed for a 16-inch viewing distance, they must lean in toward a screen which may be 20–25 inches away in order to clear the image. If the same worker is using traditional bifocals, which are designed to see the near object in the lower visual field, they must tilt their head backward and lean forward to put the viewing section of the lens into proper position to see the screen. If a VDT worker is viewing hard copy, which often is off to one side, they might need to keep moving their head back and forth to view the screen alternatively with the hard copy. This will also lead to neck discomfort.

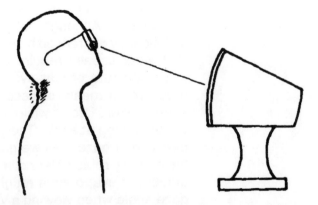

Figure 5.2 Bifocal lenses require the wearer to gaze downward in order to properly see through the reading portion of the lens. This will necessitate a head tilt if the viewing angle is too small.

Computer Vision Syndrome

These and many other situations are all too common in the office environment and cause excessive postural accommodations which lead to the symptoms of neck and back discomfort. Lie and Watten (1994) found that doing VDT work for three hours contributed not only to eye muscle fatigue but also muscle pain in the head, neck and upper back regions. Fahrback and Chapman (1990) found the highest areas of complaints was the head for heavy VDT users and the back for light VDT users. One of the main reasons for these problems is the setup of the workstation, most often the position of the monitor. All too frequently the monitor is placed either on top of the Central Processing Unit (CPU) or on a monitor stand. This places the screen in a position where the user must look either straight ahead or actually upward in their gaze.

Eye level is often determined with the user sitting 'tall'. However, in normal, upright sitting (without a visual target), Hsiao and Keyserling (1991) found that subjects tilted their head and neck an average of 13 degrees forward from the upright position. If the monitor is set to eye level, the user is presented with a choice: either assume a more erect head/neck posture than preferred or employ a gaze angle above the

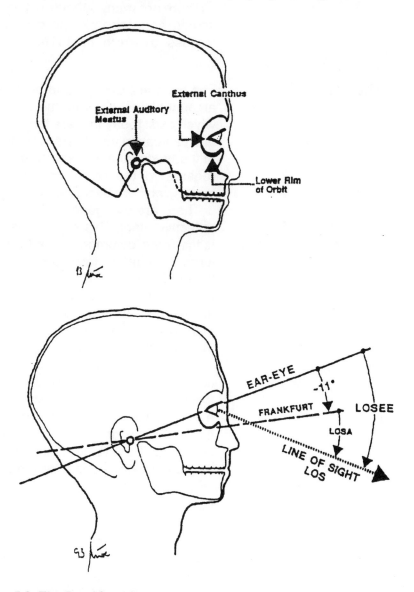

Figure 5.3 The Frankfurt Line.

reference line which passes through the right ear hole and the lowest part of the right eye socket (Frankfurt Line).

When the head-erect posture becomes tiring, users have limited possibilities for relief. One option is to tilt the head backward (extension). Hill and Kroemer (1986) found that when users in an upright seated posture were shown targets 50–100cm away, they preferred to gaze an average of 29 degrees below the Frankfurt Line. Another alternative posture available to computer users with eye-level monitors is the forward head position, in which the head remains erect while jutting forward from the trunk. Users sometimes assume a forward head posture in a counterproductive attempt to relieve muscle tension caused by contracted neck muscles (Mackinnon and Novak, 1994).

The last alternative neck posture available with an erect trunk position is flexion, or forward bending. Chaffin (1973) found that 15 degrees of sustained neck flexion for a long period (six hours with 10 minute breaks each hour) resulted in no elevated electro-muscular reading or subjective reports of discomfort. Sustained neck tilts of more than 30 degrees, however, greatly increased neck fatigue rates.

There are many schools of thought as to the 'perfect' height for the screen and each has its own supporting theories. These will be discussed in further detail in Chapter 9.

Light sensitivity

The eyes are designed to be stimulated by light and to control the amount of light entering the eyeball. There are, however, conditions which exist today that are foreign to the natural lighting environment and can cause an adverse reaction to light. The largest single factor in the workplace is glare. Glare will be discussed in more detail in regards to the remedies for the workplace but it bears some discussion here because it is a significant factor in CVS.

There are two general categories of glare: discomfort glare and reflective glare. This section discusses discomfort glare because it is the more common cause for light sensitivity. Discomfort glare is largely caused by large disparities in brightness in the field of view. It is much more desirable to eliminate bright sources of light from the field of view and strive to obtain a relatively even distribution of luminances. A person is at greater risk of experiencing discomfort glare when the source has a higher luminance and when it is closer to the point of attention.

One of the primary reasons discomfort glare is a problem for computer users is that light often leaves the overhead fluorescent fixture in a wide angle, resulting in light directly entering the worker's eyes. It is very common for the luminance of the fixture to be more than 100 times greater than that of the display screen that the worker is viewing. This is a particular problem of computer workers because they are looking horizontally in the room (assuming the screen is at 'eye level'). Bright open windows pose the same risks as overhead light fixtures.

Workers are also at risk for discomfort glare if they use a dark background display screen, resulting in a greater luminance disparity between the task and other object in the room. Other sources of large luminance disparity at the computer workstation include white paper on the desk, light colored desk surfaces and desk lamps directed toward the eyes, or which illuminate the desk area too highly.

Double vision

In Chapter 2 we discussed the process of binocular vision and how we manage to see just one image while using two eyes. This normal viewing process can be disturbed by excess use, especially when looking at a close working distance for extended periods of time. When we lose our ability to maintain the 'lock' between the eyes, they can mis-align and aim at different points in space. If both eyes keep transmitting the image back to the brain, we will experience double vision, or *diplopia* (di-PLO-pee-ah).

Double vision is a very uncomfortable and unacceptable condition for our visual system. We will most likely 'suppress' or turn off the image of one eye rather than experience the double images. When viewing a near point object, the extraocular muscles 'converge' the eyes inward toward the nose. Convergence allows the eyes to maintain the alignment of the image on the same place on both retinas.

We previously discussed the resting point of accommodation (Chapter 2) but there is also a resting point of convergence. This point varies among individuals but the average is about 100cm (Jaschinski-Krusza, 1991). Looking at objects closer than one's resting point causes strain on the muscles controlling the vergence. The closer the distance, the greater the strain (Collins, 1975). In fact, the resting point of convergence has an even greater impact on eyestrain than the resting point of accommodation. Jaschinski-Krusza (1988) measured productivity on a group of subjects which was maximum at the 100cm distance. Owens and Wolf-Kelly (1987) found that after one hour of near work, the resting points of both accommodation and vergence demonstrated an inward shift. The magnitude of the shifts depended on the positions before the near work; subjects with initial far resting points exhibited the greatest inward shifts.

Often a worker will not experience this doubling of vision while using the VDT but afterward. This is a sign that the convergence system is working, but unable to *stop* working! If the resting point of vergence is so far inward that distance objects cannot be viewed properly, they will appear to double. Fortunately, this problem is not predominant mainly due to the visual system's 'survival' instinct and suppression ability. However, it is unfortunate as well because the symptom of this type of stress may not be noticeable until it reaches an advanced stage. This is further evidence for the need for periodic eye examinations which can determine a worker's vergence abilities.

After-images and color distortion

Anyone who has had their picture taken with a camera with flash attachment has seen an after-image. It is that persistent image of the light which we still see for some time after the initial flash has gone. It is beyond the scope of this book to discuss the physiological reasons for this effect but it is normally of no consequence because it dissipates within a short time. It has, however, been reported in some cases of computer users who are looking at an excessively bright screen for an extended period of time.

Our retina is also responsible for our perception of color vision. Although we still have only a 'theory' of color vision, we have a pretty good idea of how it works. It is mediated by the cones of the retina and when they are exposed to a particular color for an extended period of time, they become 'bleached', or desensitized to that color. Since those cones are temporarily non-functional, the neighboring cones become more effective and they produce a color which is complementary to the original bleaching color. This condition is called the McCullough Effect (McCullough, 1965). For example, looking at the color green for a long time will exhibit a red (or pinkish) after-image when looking at a white surface. This has been demonstrated in almost 20% of users in a study (Seaber *et al.*, 1987) but there was no permanent damage and it could not be determined who is more likely to experience the effect. Working on a full color monitor with various colors used throughout a day will probably not create this condition.

Computer Vision Syndrome is a by-product of excessive viewing of VDT screens without regard to common sense visual hygiene. By just using some common sense and education about the visual system, the symptoms of CVS can be diminished or eliminated.

References

Chaffin, D.B. (1973) Localized Muscle Fatigue – Definition and Measurement. *Journal of Occupational Medicine*, **15 (4)** 346–354.

Collins, C.C., O'Meara, D. and Scott, A.B. (1975) Muscle Strain During Unrestrained Human Eye Movements. *Journal of Physiology*, **245**, 351–369.

Collins, M.J., Brown, B. and Bowman, K.J. (1991) Task Variables and Visual Discomfort Associated with the use of VDTs. *Optometry and Visual Science*, **68:1**, 27–33.

Dain S.J., McCarthy, A.K. and Chan-Ling, T. (1988) Symptoms in VDU Operators. *American Journal of Optometry and Physiological Optics* **6**, 162–167.

Fahrback, P.A. and Chapman, L.J. (1990) VDT Work Duration and Musculoskeletal Discomfort. *AAOHN Journal*, **38(1)**, 32–36.

Galinsky, T.L., Rosa, R., Warm, J.S. and Dember, W. (1993) Psychophysical Determinants of Stress in Sustained Attention. *Human Factors*, **35(4)**, 603–614.

Hill, S.G. and Kroemer, K.H.E. (1986) Preferred Declination of the Line of Sight. *Human Factors*, **28(2)**, 127–134.

Hsiao, H. and Keyserling, W.M. (1991) Evaluating Posture Behavior During Seated Tasks. *International Journal of Industrial Ergonomics.* **8**, 313–334.

Jaschinski-Krusza, W. (1988) Visual Strain During VDU Work: The Effect of Viewing Distance and Dark Focus. *Ergonomics*, **31(10)**, 1449–1465.

Jaschinski-Krusza, W. (1991) Eyestrain in VDU Users: Viewing Distance and the Resting Position of Ocular Muscles. *Human Factors*, **33(1)**, 69–83.

Lie, I. and Watten, R.G. (1994) VDT Work, Oculomotor Strain and Subjective Complaints: An Experimental and Clinical Study. *Ergonomics*, **37(8)**, 1419–1433.

Luberto, F., Gobba, F. and Brogha, A. (1989) Temporary myopia and subjective symptoms in video display terminal operators. *Med Lav* **80(2)**, 155–163.

Mackinnon, S.E. and Novak, C.B. (1994) Clinical Commentary: Pathogenesis of Cumulative Trauma Disorder. *Journal of Hand Surgery*. 19A, **(5)**, 873–883.

McCullough, C. (1965) Color adaptation of Edge Detectors in the Human Visual System. *Science* **149**, 1115.

Owens, D.A. and Wolf-Kelly, K. (1981) Near Work, Visual Fatigue and Variations of Oculomotor Tonus. *Investigative Ophthalmology and Visual Science*. **28**, 743–749.

Patel, S., Henderson, R., Bradley, L., Galloway, B. and Hunter, L. (1991) Effect of Visual Display Unit Use on Blink Rate and Tear Stability. *Optometry and Visual Science* **68(11)**, 888–892.

Rosenfeld, M. and Ciuffreda, K.J. (1994) Cognitive demand and transient near-work-induced myopia. *Optometry and Visual Science*, **71(6)**, 475–481.

Seaber, J.H., Fisher, B., Lockhead, G.R. and Wolbarsht, M.L. (1987) Incidence and characteristics of McCullough aftereffects following video display terminal use. *Journal of Occupational Medicine*, **29: 9**, 727–9.

Sheedy, J.E. (1992) Vision problems at Video Display Terminals: A Survey of Optometrists. *Journal of American Optometric Association*, **63**, 687–692.

Smith, M.J., Cohen, B.G.F. and Stammerjohn, L.W. (1981) An Investigation of Health Complaints and Job Stress in Video Display Operations. *Human Factors*, **23 (4)**, 387–400.

Sotoyama, M., Vilkanneva, M.B., Jonai, H. and Saito, S. (1995) Ocular surface area as an informative index of visual ergonomics. *Industrial Health*, **33:2**, 43–55.

Tsubota, K. and Nakamori, K. (1993) Dry Eyes and Video Display Terminals. *New England Journal of Medicine*, **328**, 8.

Udo, H., Tanida, H., Itani, T., Otani, T., Yokata, Y., Udo, A., Omoto, Y., Tuboya, A. and Yokoi, Y. (1991) Visual load of working with visual display terminal – introduction of VDT to newspaper editing and visual effect. *Journal Human Ergonomics* (Tokyo), **20 (2)**, 109–121.

Watten, R.G. and Lie, I. (1992) Time factors in VDT-induced myopia and visual fatigue: an experimental study. *Journal Human Ergonomics* **21(1)**: 13–20.

Wiggins, N.P. and Daum, K.M. (1991) Visual Discomfort and Astigmatic Refractive Errors in VDT Use. *Journal of American Optometric Association* **62 (9)**, 680–684.

Yaginuma, Y., Yamada, H. and Nagai, H. (1990) Study of the Relationship Between Lacrimation and Blink in VDT Work. *Ergonomics* **33(6)**, 799–809.

Yamamoto, S. (1987) Visual, musculoskeletal and neuropsychological health complaints of workers using video display terminal and an occupational health guideline. *Japan Journal Ophthalmology*, **31:1**, 171–183.

Chapter 6

Vision examinations

Introduction	This chapter is not designed to teach you how to conduct an eye examination nor to second-guess your doctor. It is here to illustrate the complexities that the VDT has introduced into the examination procedure. You will learn how to actually assist your doctor in completing a more comprehensive and appropriate eye examination. The first step is to make sure that you know some basic information about eye care professionals. There are typically three 'Os' which people tend to confuse when talking about eye care professionals. Let's take a look at these professions and see what distinguishes them.
Ophthalmologist	An ophthalmologist (of-thal-MAHL-oh-jist) is a physician (MD) who specializes in the comprehensive care of the eyes and visual system in the prevention of eye disease and injury. The ophthalmologist has completed one year of internship and three or more years of specialized medical and surgical training and experience in eye care. They are qualified to diagnose, treat, and manage all eye and visual system problems, and are licensed by a state regulatory board to practice medicine and surgery.
Optometrist	An optometrist (op-TAHM-e-trist) is a doctor of optometry (OD). Doctors of Optometry are independent primary health care providers

who examine, diagnose, treat and manage diseases and disorders of the visual system, the eye and associated structures as well as diagnose related systemic conditions. The optometry curriculum is a four-year course following undergraduate study and often includes internships at various clinical facilities. Optometrists are licensed by individual state boards of optometry. The essential difference between the two doctors is that ophthalmologists perform surgery when necessary.

Behavioral optometry is a subspecialty of optometry which includes developmental and functional optometrists – all of whom deal in depth in preventing or correcting visual problems, charting visual development and enhancing visual performance. Behavioral optometrists have found that the visual care needs of VDT operators differ from the general population and perform more extensive testing in determining the proper treatment plan.

Optician

An optician (op-TISH-an) is a technician who is trained to fill prescriptions for lenses written by optometrists or ophthalmologists. They are trained to make glasses, fit eyeglass lenses into frames and adjust glasses frames to a person's face. In some states, opticians are allowed to do the fitting of contact lenses.

The usual course for opticians is an associate college degree, which is normally a two-year program. Most states license opticians and require continuing education. There is an American Board of Opticianry, which certifies opticians; this certification may or may not be accepted by a particular state as qualifying an optician to practise. At this time, requirements vary a great deal from state to state.

All three of these professionals have an important job to do in maintaining your visual well-being.

The traditional eye examination

You might think that an eye examination is just one standard set of tests which doctors automatically perform to determine what prescription is suited for optimum vision. Wouldn't it be nice if that were true! There is, however, much subjectivity in an eye examination and, in fact, there are a number of different types of examinations which might be performed depending on the nature of the symptoms presented and the expertise of the doctor performing the examination.

There are many components of an eye examination but there are some basic areas which all examinations must evaluate. These consists of eye tests which assure that you see well and that the eyes are healthy. Unfortunately, many doctors might stop at these tests and send you on your way with a reassuring '... you're seeing 20/20' statement. For example, the traditional medical model of vision (subscribed to generally by the ophthalmologist) is one in which an eye is disease-free and has clear distant vision. Of course, if there is a disease process then further testing would be conducted to determine the cause and remediation of the problem. However, a thorough eye examination must go further, especially when considering the VDT user.

You'll recall from Chapter 2 that our traditional far-point for testing distance vision is 20 feet. This is the distance at which light enters the eye as if it were coming from an infinite distance, so 20 feet is all that is needed. This is a valid way to test the visual sharpness, or 'acuity' of the eye. Theoretically, the eye should be in a relaxed state and seeing clearly at this distance. However, we now know this not to be the case because we are dealing with an active muscular system which reacts to many biological, psychological and environmental conditions. Over the past 50 years, many doctors (generally behavioral optometrists) who have learned about these factors, have tested at a near point of vision as well. This testing distance has been 16 inches because it was found that this is the point where we comfortably hold reading material. A testing protocol was established and a series of testing procedures were developed so doctors had standards by which to measure the patients' visual performance. This series of tests served the doctors and patients well, especially where children's school performance was concerned. The doctors simply set up a near-point target at 16 inches and ran these tests in a similar manner to every patient. It actually was a good system and worked rather well. But then came the computer!

The occupational vision examination

The unique characteristics and high visual demand of VDT work make eye and vision problems the most frequent health-related problem experienced by VDT users. Uncorrected vision conditions, poor VDT design and workplace ergonomics, and a highly demanding visual task can all contribute to the development of visual symptoms. The American Optometric Association recommends that all VDT operators obtain a comprehensive eye/vision examination prior to or soon after beginning VDT work and periodically thereafter. The examination should include:

1. A general systemic and ocular health history.
2. A specific patient history relating to VDT use (see next section).
3. Measurement of unaided and aided visual acuity at distance and appropriate near distances.
4. Evaluation of internal and external eye health.
5. Refraction at distance and near working distances.
6. Assessment of eye focusing abilities.
7. Evaluation of eye coordination and eye movement skills.

Depending on the history, signs and symptoms presented or the results of the initial examination procedures, additional testing may be indicated to effectively evaluate and diagnose specific eye or vision problems. At the completion of the examination, you should receive a review and discussion of the finding and recommendations for treatment. Treatment may include the design and prescription of a lens correction specifically for VDT work, the utilization of optometric vision therapy for the correction of binocular vision and accommodative dysfunction interfering with VDT work, and/or guidance relating to preventive eye care, lighting and workstation design.

VDT and vision questionnaire

The key to addressing all of the pertinent issues relating to VDT work is knowledge of the working environment. This is one of the most difficult areas for doctors to address because they are performing examinations in a room at their office which has controlled environmental conditions. If they are to correctly determine what prescription to apply, they must be able to evaluate the patient's working environmental, as well as the visual, conditions.

One of the easiest ways to accomplish this is with a VDT and vision questionnaire (see Appendix A). This survey is designed to give the doctor critical information regarding the environmental condition in which the patient is working and what uncontrolled factors should be considered. This questionnaire has many forms and the enclosed version is certainly not the last word but should cover all the areas of interest to the doctor. The questionnaire may appear very thorough but there may be other signs or symptoms which arise that do not appear here. List them at the end of the questionnaire and remember to mention them to your doctor.

Vision screenings

An intermediary method of obtaining cursory information about visual abilities on the job is through a vision screening. A screening is a series of tests which approximate the visual abilities of the individual so it can be decided if further, more in-depth evaluation is required. There are three general categories of screening procedures currently available: professional testing, stereoscopic testing and computer-generated testing.

The professional vision screening is a valuable method for finding vision problems in an office environment. It involves the services of a qualified eye care professional – usually an optometrist or ophthalmologist – and any support staff which they may require. Often there is a test workstation set up where they can conduct the tests on individual workers who are appointed to come there at certain times. The doctor performs the series of various tests and coordinates the results with a questionnaire to determine if a full examination is warranted. If possible, the doctor can also visit the employee's own workstation to evaluate environmental factors.

The advantage of this type of screening is that there is a (hopefully) qualified eye care professional who is recording valuable information about the visual system of the worker. If the same doctor is performing the full vision examination, then they will have a distinct advantage in determining the proper recommendation. The disadvantage of this is often the cost of having a doctor and staff member taking time to perform all of these tests. Additionally, the worker must take the time off from their workday to attend the session.

A second, and the most common form of vision screening, is the stereoscopic method. Most people are familiar with the 'box' into which they gaze to read letters and view targets. These conventional vision screeners have been used for many years, often for drivers' testing, and use targets which are simulated to various distances, e.g. 20 feet or 16 inches. The range of visual functions assessed by these instruments and the degree of automation varies between models but most instruments permit an adequate assessment of visual function.

Although these instruments are useful for general screening, their suitability for screening VDT users is questionable for the following reasons:

1. The test conditions are different from normal VDT viewing conditions. For example, the test targets bear little resemblance to

objects displayed on a screen (in terms of luminance, flicker, contrast, etc.) and the field size and angle of gaze is constrained by the instrument. Also, the fixed 'intermediate' distance may be different to a user's actual screen distance. Furthermore, the act of looking into an optical instrument may induce 'instrument' accommodation and convergence which will shed further doubt on the validity of the results.

2. Many of the tests performed by the current generation of vision screeners have no direct bearing on the ability to use a display screen efficiently and comfortably.

3. Interpretation of the results of a screening test requires a certain level of optometric knowledge combined with detailed information about the user, their working environment and work practices.

The third type of screening is a computer-based system which presents a questionnaire and performs a series of vision tests directly on the user's own display screen. There are just a few of these types of programs available (just one in the United States as of this writing) and they appear to be the most effective form of screening available. The purpose of the program is to identify the following groups of subjects: (1) Those who are not experiencing any problems with their eyes and have no measurable visual defects; (2) Those who are not experiencing any problems with their eyes but have a measurable visual defect; (3) Those who are experiencing problems with their eyes but have no measurable defects; and (4) Those who are experiencing visual problems and have measurable eye defects.

The program, called *The Eye-CEE System for VDT Users*® is a Windows®-based software which can be administered by a Safety & Health administrator or self-tested on a local network. The on-line questionnaire asks the user a series of questions about their working conditions and complaints they might notice. It then carries out a series of specific vision tests. The results of the tests and answers to the questionnaire are compiled and reports are optionally generated for the employee, the administrator, the employer and for the eye care professional. In addition to the basic screening, the program provides a number of administrative functions and can be designed to print specific reports, make ergonomic recommendations, or tailored to meet any specific workplace needs. Because this program is conducted right on the display screen of the employee, it provides the most accurate information on visual function in the workplace. In addition, it is priced very reasonably, based on the number of users tested. Further information on this program is available from the resource list in the back of this book.

Whichever method is chosen for evaluation, it is highly recommended that the environmental considerations be addressed as critically as possible. Our vision can only be as accurate as our environment allows it to be.

Chapter 7
Vision in industry

Introduction

While this book has been directed primarily toward the use of the VDT, we should remember that there is still a large portion of our workforce who deal with industrial production. There have been many articles written and discussions about eye safety as the key element in regard to industrial vision. However, there is more to maintaining proper vision in the industrial setting. The arena of functional vision has found a niche here as well and it is an issue which should be rightfully addressed. This chapter will discuss both eye safety and functional vision and show how they can be maximized in the workplace.

The cost of eye injuries

There are significant costs involved with occupational illness and injury. What most employers don't realize, however, is that there are two distinct categories of cost: the obvious cost and the hidden cost. The obvious costs of accidents are the medical and compensation expenses. The hidden costs are the costs associated with training replacement workers, property and equipment damage, missed deadlines, production delays, investigation time, overtime and downtime costs and reduced employee attendance and morale. Research has shown that for every dollar spent on workers' compensation costs, at least four dollars are lost on hidden, often unrecorded costs.

A total of 6.8 million injuries and illnesses was reported in private industry workplaces during 1994, resulting in a rate of 8.4 cases for every 100 full time workers. Of this total of 6.8 million, nearly 6.3 million were injuries that resulted in either lost work time, medical treatment other than first aid, loss of consciousness, restriction of work or motion, or transfer to another job. The remainder of these private industry cases (about 515,000) were work-related illnesses. Employers and employees in private industry and state and local governments spent $258.5 billion for health care plans during 1992. Employer expenditures for workplace-based health care plans ($221.4 billion) were nearly six times those of employees ($37.2 billion). These numbers are expected to rise with every reporting year since 1992 (Bureau of Labor Statistics, 1993).

The costs associated with some eye and vision injuries can be estimated because of the need for treatment and workers' compensation costs, both of which are obvious costs. Within a specific workplace, the amount paid for eye injuries can be significant, especially if an eye is lost. The direct costs of a single employee losing one eye range from about $40,000 to $115,000 (Thackray, 1982). Workers' compensation laws have the loss of one eye as a scheduled benefit ranging form $5,699 to $157,685, depending on the state (US Chamber of Commerce, 1989).

Prevent Blindness America estimates that 300,000 disabling eye injuries as far back as 1982 cost business and industry $330 million in lost production time, medical bills, and compensation. However, they also suggest that 90% of eye injuries are preventable. According to the Bureau of Labor Statistics, 60% of workers who experienced impact or chemical burn eye injuries accident were not wearing eye protection. Most injured workers were on the job doing their normal activities and received many different types of eye injuries. Of the objects striking the eye, almost 60% were less than 0.5mm, which is smaller than a pinhead. Two thirds of the objects were travelling faster than a hand-thrown object. Chemical contact was responsible for 20% of the injuries. Of the remaining 40% of workers who had been wearing eye protection, 40–80% of them were not wearing safety glasses (US Dept. of Labor, 1980).

In a study in California in 1989, the major causes of eye claims for workers' compensation were scratches and abrasions (66.8%), diseases of the eye (13.6%), burns and scalds (7.0%), cuts, lacerations, and punctures (5.1%), radiation effects (5%), infective or parasitic diseases (1.6%) and other (0.9%). Unlike the widespread awareness of medical costs in industry, the high costs associated with untreated vision disorders are unrecognized and not easily quantifiable. These costs are found in reduced productivity of workers and unnecessarily high rates of spoiled or second-class products. These costs also include the costs of accidents and co-worker injuries that could have been prevented if vision disorders had been treated. No business or industry accounts for the costs of untreated vision disorders, and therefore they are not recognized as a significant problem.

Industrial vision requirements

Much progress has been made over the past several years in attempting to reduce the illness/injury statistics. Yet, there is still one important aspect of this work which is all too often neglected – the visual efficiency of the workers themselves. It has been established that 25–40% of workers have vision below the standard required for their occupation (Grundy, 1981). In many cases, this is a result of uncorrected or insufficiently corrected visual defects which can affect both the well being of the worker and their efficiency, productivity and safety. Each worker must possess adequate vision for the work he or she is required to undertake. Some work requires the highest visual standards, whereas other jobs can be performed adequately by persons with relatively poor vision.

There has been considerable amount of research supporting the relationship between accidents and defective vision (Koven, 1947; Tiffin and Wirt, 1945; Steward and Cole, 1989). Vision screening carried out in a large steel works showed that employees who did not come up to the visual standard were found to have experienced, on average, 20% more accidents than those who were visually

Table 7.1 The visual requirements of several different occupations

Visual Skills	VDT Operator	Clerical/ Admin.	Inspection	Mobile Equipment	Machine Operator	Mechanic	Unskilled Labor
Locating, Scanning, Tracking	X	X	X	X	X	X	X
Peripheral Vision	X	X	X	X	X	X	X
Depth Perception			X	X	X	X	
Color Vision			X	X	X	X	
Glare Recovery	X		X	X	X	X	
Eye-Hand Coordination	X		X	X	X	X	
Focusing Range/Speed	X	X	X	X	X	X	
Far: Central Vision		X		X	X	X	X
Far: Equal Focus				X			
Far: Sustained Focus				X			
Ease of Single Vision				X			
Near: Central Vision	X	X	X		X	X	
Near: Equal Focus	X	X	X		X	X	
Near: Sustained Focus	X	X	X		X	X	
Ease of Single Vision	X	X	X		X	X	
Sustained Single Vision	X	X	X		X	X	

Copyright ©1983 by Occupational Vision Services, Inc. Burbank, CA. with permission.

efficient. It is only logical to assume that a worker can't do his/her job with confidence, accuracy and efficiency if they can't see it properly. Table 7.1 offers some various occupation categories and the different visual requirements for each.

Some other considerations in the workplace are factors for safety and the visual environment. For example, there is illumination contrast: great differences in illumination between different parts of the work area, such as between store rooms and a customer-service area, can be dangerous if employees must move frequently between these areas. The eyes take only a few minutes to adapt from dim light to bright light, but they take up to 30 minutes to adapt completely to dim light after being in bright light. The danger of errors and accidents is increased during this adaptation period (for instance, while getting something from a dimly-lit storage area after being in a brightly-lit room). All work areas that are in constant use should have nearly uniform illumination. There is also the concern of contrasting colors: the more contrast there is between an object and its background, the easier it is to perceive the object. Therefore, contrasting colors can be used to enhance performance and decrease errors and accidents. The moving parts of machines, the edges of steps, hot pipes and so on can be made much more easily visible by having them painted with a color that contrasts sharply with the surrounding color.

To obtain an effective eye examination which addresses industrial vision concerns, there is an excellent questionnaire which should be duplicated, filled out prior to the exam, and taken to the doctor (see Appendix B). As with the VDT questionnaire, the doctor will be able to use this information to better conduct the examination, as well as prescribe the appropriate vision correction, if needed.

Industrial eye protection

No discussion of vision care in the industrial setting would be complete without a review of the basics of eye safety. Prevent Blindness America (1982) estimates that about one million vision impairments are due to eye injury. Eye injury in the workplace is second only to the home where accidents happen most frequently.

The U.S. Occupational Safety and Health Administration (OSHA) requires that all industrial eye and face protectors meet the requirements of the ANSI Z87.1 standard (discussed in detail in Chapter 12). OSHA requires employers to assess the workplace to determine if hazards are present that necessitate the use of personal protective equipment (PPE). Employers must not only determine which jobs require safety glasses, they must provide a written certification that the workplace has been so evaluated. This is called a 'certification of hazard assessment'. You can have a safe operation, furnish employees with the best in PPE, and still not be in compliance if you don't have a written certification of hazard assessment in the proper form. OSHA also requires that employees be trained in: (a) when PPE is necessary; (b) what PPE is necessary; (c) how to put it on, off, adjust, and wear PPE; (d) the limitation of the PPE; and (e) the proper care, maintenance, useful life, and disposal of the PPE. This requires a written certification that contains the name of each employee trained and the date of training, and identifies the subject of the certification. Prescription industrial safety eye wear requires that both lenses and frames meet the ANSI Z87.1-1989 standard.

Industrial safety lenses in a dress safety frame or dress safety lenses in an industrial safety frame do **not** meet the standard and should never be worn. The following is a summary of the requirements for all industrial safety prescription lenses:

1. All lenses must be able to withstand the impact of a 1-inch steel ball dropped 50 inches onto the lens front surface (the industrial safety drop-ball test; see below).

2. All lenses must have a minimum thickness of 3mm. Lenses of power greater than +3.00 diopters in the most plus meridian can have a minimum thickness of 2.5mm.

3. All lenses must be permanently marked with the monogram or trademark of the optical laboratory that edges the lenses. This provides proof to a manager or supervisor in the workplace that an employee's lenses are truly industrial safety lenses.

4. In general, the standard does not prohibit the use of tints for industrial safety lenses. This would include, for example, sunglass tints and light tints for comfort or cosmetic purposes. However, tints must be prescribed with knowledge of the patients' lighting and working conditions. For example, lenses of low transmittance (e.g. sunglasses) should not be worn indoors, and lightly tinted lenses for indoor use should be prescribed only if the workplace has adequate lighting or if glare is a problem. Sunglasses should not be worn by the operator of a vehicle with a tinted windshield, nor should they be worn for driving at night. A tinted lens is classified as a protective filter (for use in welding and other potentially hazardous situations) only if the tint meets the ANSI Z87.1 standards for minimum and maximum visible light transmittance and maximum ultraviolet (UV) and infrared (IR) radiation transmittance. The tints used for dress safety prescription lenses or sunglasses usually will not meet these standards.

Table 7.2 can be used as a guide when considering lenses for use near ultra-violet light sources.

5. Requirements for prescription power accuracy are those of ANSI Z80.1-1987. Industrial safety spectacle frames for use in prescription eye wear must meet a number of performance standards, including high mass–low velocity and low mass–high velocity impact resistance standards, flammability standards, and corrosion standards.

Polycarbonate is the best lens material to use in prescription industrial safety eye wear. The decreased scratch resistance of coated polycarbonate relative to crown glass should not be

Table 7.2 Ultraviolet absorption of various types of lens materials

Material	Index of Refraction	Absorption UVC*	Absorption UVB	Absorption UVA
Glass	1.523	100%	80%	15%
CR-39 Plastic	1.498	100%	100%	85%
Hi-index Plastic	1.43/1.56/1.60	100%	100%	90–100%
Hi-index Glass	1.60/1.70/1.80	100%	100%	80%
Polycarbonate	1.586	100%	100%	95–100%

* This range of UV light does not normally reach the earth's surface.

considered reason for using glass unless the work environment provides very severe conditions. Polycarbonate can be a problem in cold, dusty work environments because static charges cause dust to cling to the lenses.

OSHA is considering a revision of its general industry standards for eye and face protectors to include issues not specifically addressed by the ANSI Z87.1 standard. One important potential revision is a requirement for side shields on all safety spectacles. This revision is based upon a report by the Bureau of Labor Statistics (1992) that 94% of occupational eye injuries studied were the result of projectiles reaching the eye from the unprotected sides, tops or bottoms of the protector. OSHA is also considering changes in the way eye and face protectors are labeled and a requirement for third-party certification of protectors.

Ophthalmic lenses may be divided into two types for purposes of impact resistance: dress safety lenses worn every day and industrial safety lenses worn for occupational or educational eye protection. The standards for dress lenses are regulated in the United States by the Food and Drug Administration (FDA). Effective from 1 January 1972, the FDA made it illegal for ophthalmic practitioners to prescribe dress safety lenses that are not impact resistant. The impact resistance standard test is the 'drop ball' test, whereby a lens must be able to withstand the impact of a $\frac{5}{8}''$ steel ball weighing approximately 0.56 oz. dropped 50 inches onto the front surface of the lens. The test method is similar to that described for dress safety lenses in ANSI Z80.1-1987.

Many of the concerns of lens protection have been eliminated with the introduction in 1985 of the polycarbonate material. Polycarbonate is a plastic which has a higher density than conventional plastic lens material, rendering it essentially invulnerable to breakage. Because of its unique properties, it can also be made thinner and lighter than conventional plastic while maintaining its unbreakable status. Polycarbonate is the only material that can adequately resist the high energy impact of both large and small objects. Whenever eye protection is a major concern, only polycarbonate provides adequate protection.

In summary, the optimum eye wear for industrial use should address the following concerns: it should satisfy the prescription needs of the wearer, be comfortable, be of industrial strength, allow for adequate peripheral vision, be fog resistant, protect against excessive ultra-violet light, be cost effective and be readily available.

Contact lens concerns

In risk management, it is necessary to balance risk with benefits and to differentiate a perceived risk from the actual risk. Obsessive and unrealistic risk avoidance with its accompanying over-regulation and bans may reduce quality of life and productivity while contributing little to safety. Because contact lenses in certain environments or situations may produce adverse ocular effects, it is tempting to assume that a contact lens wearer is a greater risk in a hazardous environment because of presumed effects. Often it has been the general policy of a company simply to ban contact lenses in any situation in which there is a perceived or actual risk to the eye without regard to all factors involved.

There may be many reasons why individuals choose not to remove their contact lenses in the workplace or in an environment perceived to contraindicate contact lens wear. These may be optical, therapeutic, hygienic or cosmetic, or it may be simply that they do not accept or recognize the potential risk.

When considering the advisability of wearing contact lenses in a given work environment, a number of questions must be addressed:

1. Is there an actual ocular hazard?
2. Does the wearing of a contact lens place the eye at greater risk than a naked eye?
3. Does the removal of a contact lens increase the risk to the eye or increase its susceptibility to insult?
4. Is the risk different for various contact lens designs and materials?
5. Are there associated risks for the contact lens wearer who removes lenses?
6. Do contact lenses decrease other safety strategies?

If a potential hazard in wearing contact lenses in a specific workplace is identified, it is essential to consider the effects of the various risk factors encountered, both on the eye and on the individual who is not wearing contact lenses. Once these are determined, the situation may be evaluated theoretically using the known physical parameters and physiological effects of the contact lens. There may or may not be laboratory studies, epidemiological data, or well-documented case reports to support or refute the theoretical conclusions. When an incident involving contact lens wear is assessed, the pitfall is to assume that because the victim was wearing contact lenses, they were the causative or a contributory factor.

The state of technology involved with contact lenses dictates that they are a safe and effective method of vision correction. Careful evaluation of the workplace situation is warranted to determine if there are any contraindications to successful contact lens wearing on the job. It will most likely be found that contact lenses can be worn in many of the same viewing and working situations as glasses or no visual correction.

References

Bureau of Labor Statistics (1993) Expenditures for health care plans by employers and employees, 1992. Press Release.

Division of Labor Statistics and Research (1989) California Work Injuries and Illnesses. San Francisco, CA: Dept. of Industrial Relations.

Grundy, J.W. (1981) Eyes Geared for the Job. *Optical World*, Vol. V, 4–5.

Koven, A.L. (1947) Right eyes for the right job. Transcript of the American Academy of Ophthalmology and Otolaryngology, **50**, 46.

Prevent Blindness America (1982) A Guide for Controlling Eye Injuries in Industry. Chicago: PBA.

Steward, J.M. and Cole, B.L. (1989) What do color vision defectives say about everyday tasks? *Optometry and Visual Science*, **66**, 935–938.

Thackray, J. (1982) The High Cost of Workplace Eye Trauma: The 90% Avoidable Injury. *Sightsaving Magazine*, **51(1)**, 19–22.

Tiffin, J. and Wirt, S.E. (1945) Determining visual standards for industrial jobs by statistical methods. Transcript of the American Academy of Ophthalmology and Otolaryngology, **50**, 72.

US Chamber of Commerce (1989) *Analysis of Workers' Compensation Laws*. Washington, DC, US Chamber of Commerce.

US Department of Labor (1980) Accidents Involving Eye Injuries. Report 597. Washington, DC: Bureau of Labor Statistics.

Chapter 8
Computing for the visually impaired

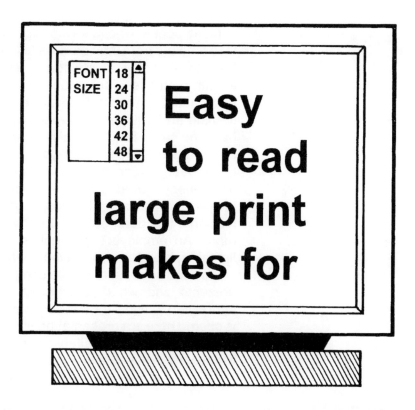

Introduction

If you have been reading this book from the beginning without any difficulty, then your eyes must be working rather well – congratulations. However, that isn't the case for a great number of people who have a visual impairment which is not correctable with traditional treatments. Whether from a genetic defect, a disease process, secondary to an accident or a progressive degeneration, visual impairments are a serious problem for millions of people. However, despite their disadvantage, these people can still be very productive members of our society and should still be offered the opportunity to contribute.

'Low' vision, as it is often called, implies some uncorrectible reduction in visual acuity and/or visual field. There is no clear consensus of what constitutes low vision or visual impairment but a person should be considered visually impaired when his or her vision is not adequate to meet their individual needs. The term 'legal blindness' is bandied about quite often but it is grossly misunderstood. State and Federal laws dictate that a person is legally blind when their *best* corrected vision is 20/200 or less, or their visual field (peripheral vision) is limited to a maximum of 20 degrees. It is estimated that more than 10 million people in the United States are visually impaired.

Since the advent of the personal computer and its widespread proliferation, people with disabilities have been concerned with access to its benefits. For the general population, the promise of this new technology was to bring convenience and order to our busy, cluttered lives. For people with disabilities the potential was much greater. People who previously needed human assistance to read a document, write a letter or play a game could now do so independently. The infinite patience of the computer would allow discovery and exploration at one's own pace.

Many people who have lost significant levels of visual functioning take time out from their careers to learn alternative skills which allow them to live and work effectively. If a person who is blind or visually impaired applies for a job in your company, the résumé and application will indicate experience and skills. With appropriate training and equipment, people who are blind or visually impaired have the same range of abilities as anyone else. There are no 'jobs for blind people'. To broaden your thinking, consider that blind people have been successful as artists, machinists, auto mechanics, masseuses, boat builders, mayors, computer programmers, lawyers, musicians, fashion models, production workers, professional teachers, story tellers, word processing specialists, and more. An employer's perception of inability is often the biggest limitation faced by people who are blind.

We will discuss some of the complications that a visual impairment presents and what adjustments and considerations should be addressed to allow for a successful working situation. Much of what is offered here is a presentation of the various hardware systems and software programs available and where to obtain them. Complete listings are presented in Appendices C and D at the end of this book.

Working with visually impaired employees

Employees who are blind or visually impaired need the same introduction to a job and initial training as sighted colleagues. Providing orientation to the company and the job and asking the employee their preferences on instruction – verbal, written or hands-on – makes job adaptation easier. Allow the employee to organize the work area for greatest efficiency, even if it means organizing it differently than in the past. You may find that the new design would make other employees more productive as well. Make sure that the workstation is adequately equipped and that the employee knows where to get replacement supplies.

A supervisor's responsibility is to establish an atmosphere of quality and productivity. Appropriate on-going training is crucial. Assume that an employee who is blind or visually impaired has the same career aspirations as other employees and provide training, ensuring that materials are accessible to the person's visual limitations. Supervise as you would for any other employee. Be sure that all employees understand performance expectation. Provide praise and constructive feedback. If a performance problem arises, deal with it openly. Do not automatically assume that it is disability-related. Do not avoid giving feedback. All employees want to know when they are performing well and when they need to do things differently. Occasionally, a supervisor will not provide feedback until a major problem arises, then dismiss the employee. This is not an appropriate supervision technique. Conduct performance appraisals, using the same criteria as with other employees. If a supervisor has provided feedback throughout the year, an annual performance appraisal should contain no surprises. Review any adaptive equipment to ensure that it is still working well. Determine if it needs upgrading to keep the employee competitive. Always promote qualified people who are blind or visually impaired using the same criteria as promotion for others.

There are two key factors which must be addressed when considering the ergonomics of a workstation for the visually impaired. Those factors are flexibility and magnification. Flexibility is of primary importance due to the likelihood of postural extremes the worker must achieve to view the VDT adequately. Depending on what the visual condition is, the worker may need to decrease their distance to the screen, view the screen at a particular angle, maintain a certain head/trunk posture or other positions. Workstations that cannot adapt to the special needs of these workers are not desirable and will not allow for a productive worker. Magnification can be achieved in two ways: by increasing the physical size of the image or by moving the viewer closer to the object. Simply moving closer to the screen is often not acceptable due to environmental restrictions or other obstacles. There are magnification programs which allow for increasing the size of the image on the screen. The following discussion will include a number of programs which increase the image size that may work well for many low vision users.

How blind persons use computers

Because blindness is not an 'all or none' condition, it is very difficult to categorically describe how different people with different levels of visual impairments adjust to their workstations. Not everyone can use computers in their standard form. If you are not able to use a keyboard or mouse, you cannot input information to the computer. If you cannot see a display screen, you cannot use the information the computer has output. Adaptive aids have been developed to replace or augment input and output devices. Input aids such as simple switches (a link to a picture of a simple on/off switch) or keyguards (a link to a picture of a keyboard template for guiding one's hands) to sophisticated speech recognition (a link to a picture of a microphone providing voice input) systems that let you speak to your computer have made access possible for many people with physical disabilities. On the output side, screen reader (link to picture of computer with a speaker for speech output) systems have been developed for people with visual impairments, that can speak what the computer displays. Computer manufacturers and developers of operating systems are now building-in access solutions or are providing them free of charge. Users of Apple Macintosh, Microsoft Windows and X-Windows systems now have better control of many input and output features.

An ongoing concern is whether new software applications will work with current technologies. The growing popularity of Graphical User Interfaces (GUIs) and windowing operating systems is an example of such a problem. When icons are used to replace text, screen readers cannot read them. In addition, application windows may cover up on-screen keyboards. These interfaces are not working with many current assistive devices and software, yet the alternative command line interface is becoming less common. It is not a significant concern if it is simply a matter of text enlargement, which may be adequate for some visual impairments. However, for the more severe conditions, special programs and equipment are required. A listing of some alternative systems follows.

Speech Output Totally blind persons usually use computers with speech output devices. Actually, there are two components: a physical speech synthesizer and a screen access program. The speech synthesizer is a circuit card or an external box that attaches to a port on the computer. Speech synthesizers cost from $300 to $2000. Most screen access programs for MS-DOS cost about $500. Most screen access programs for Windows cost about $700. One of the most economical packages is The DoubleTalk speech synthesizer with the ASAP screen access program. This combination costs $795 and is available from Raised Dot Computing. Other vendors include: Aicom (speech synthesizers) and Berkeley Systems (outSPOKEN-screen access from Windows and Macintosh).

Braille Access Braille access means using a Braille device to access the interaction with the computer. Braille is a different modality to get access to computers, compared to speech output. Braille is read by the fingers and is produced through mechanical typewriters or through a QWERTY keyboard using a PC. Each module of a transitory Braille display consists of six or eight pins. The electromagnetic technology makes a little pin move upwards or downwards to create a Braille pattern. Braille displays, like speech-based screen-readers, implement tracking of the cursor, reviewing the screen and routing. Unlike speech output, the user can effectively and accurately recognize each character while reading, and he may point at it for routing. Four generations of Braille displays exist, each providing more features to work efficiently and accurately with standard application programs.

Braille output requires a Braille embosser and a Braille translation program. Braille embossers produce hard copy Braille output from a computer. Embossers are used in conjunction with Braille translators, which control formatting and type of Braille produced. Braille embossers print from 20–50 characters per second. There are a variety of embossing machines on the market. These vary in cost from the Braille Blazer (under $2000) up to $4000 or more. Braille translation programs cost about $500. The two most prominent programs are MegaDots and Duxbury. The best variety of Braille output devices for Braille access can be found at HumanWare. The Braille Lite from Blazie Engineering can be used for Braille access.

Portable Notetakers Notetakers contain up to 640K of memory, operate comfortably on a desk or in your lap, can store more than 1000 pages, and have built-in speech synthesizers that clearly speak text that is stored in memory. Notes can be edited, stored in text files, sent to a printer, downloaded to a computer, or uploaded from a computer. Notetakers are available with a standard seven-key Braille keyboard or a standard keyboard. Several firms sell portable notetakers, such as the Braille 'N Speak (voice output), the Braille Lite (voice and Braille) (Blaise) or Braille Mate (TeleSensory).

Optical Character Recognition Optical Character Recognition (OCR) is a popular technology for blind users and for those transcribing Braille for the blind. OCR software can recognize typeset letters, numbers and other characters and send them to the computer as standard computer codes. OCR software can be trained to recognize various print formats such as bold, underlined or italic. Once the computer has the text, it can output it to speech synthesizers, Braille translation software or word processing programs. A tape recorder can be connected to the speech synthesizer to create a portable audio copy of the scanned text. Most scanners are set up to look at a page at a time.

Magnification Print enlargers magnify the contents of a computer screen or, by using closed circuit television (CCTV), enlarge printed material, hand-written information, diagrams, and even wired circuitry. Images can be enlarged 4–60 times. The user controls print size, color contrast and color selection. There are some programs included as part of the Macintosh or Windows systems software programs.

Keyboard Access Visually-impaired users generally do not need a modified keyboard. After all, a goal of touch typing is to avoid looking at the keyboard. Blind users just have a rough time cheating.

Low vision eye wear

Traditionally, eye wear for the low-vision patient has been designed for specific uses. These products include hand-held magnifiers, loupes, stand magnifiers, pinhole spectacles, high-power reading glasses, telescopic lenses, contact lens-telescopic system, and microscopic spectacle lenses. While these may be necessary to maximize the visual abilities of the visually-impaired user, they may or may not be suitable for VDT viewing.

The first consideration to address is manipulation. Any system which requires the use of one or both hands will further complicate the computer user because of the (usual) requirement of manual input into the system. Secondly, the lighting of the surrounding area may dictate unusual requirements which will actually make the VDT screen more difficult to see. Various lens tints also fall into this category because they can enhance some visual images but wash out others.

In consulting with your eye care professional, be sure to address the use of the VDT system and balance your general viewing habits with those of the display screen.

The Internet

As more people come to depend on computers, the need for more software and information becomes even greater. These are the resources that a computer needs to be of any use. People have also found a need to distribute information electronically. From disks to electronic bulletin board systems to the Internet and now the World Wide Web, more information is moving ever faster. This is the age of networked information. It is changing the way we do business, the way we educate our children and even the way we spend our leisure time. Once again, people with disabilities are concerned that this new technology will create barriers to those who stand to benefit greatly from its potential.

Networked information must be interfaced to the user through hardware and software systems. It is these interfaces that can potentially cause barriers to people with disabilities. In general, command-line interfaces to the Internet will work with current assistive technologies. For example, many text-based software packages exist for email, ftp and gopher servers that work well with screen readers and alternative keyboards.

The World Wide Web is a network of servers based on Hyper Text Transfer Protocol (HTTP). This is an information architecture that links multimedia information (text, image and sound). Client software can be as simple as a Line Mode Browser or as sophisticated as hyper media browsers, such as Netscape or Explorer. For people without full access to the Internet, there is a retrieval method using Agora, the WWW email browser. LYNX, developed at the University of Kansas, is a very popular browser due to its simple but powerful full page, text only interface. Because information on the Web can be represented in text, image or sound, one might think there would be more opportunity for people with disabilities to access information. However, if a document relies exclusively on one method to represent data, it will likely not be accessible to someone. Representing information in each of the methods does make for a more accessible document and a richer one at the same time.

A search of the Web should start at *http://www.webable.com/webcateg.html* for a very extensive listing of disability-related web sites dedicated to the blind and visually impaired.

Resources

Appendices C and D include extensive listings of companies and organizations which distribute catalogs, sell products and other useful information for blind and visually-impaired persons. These lists are not intended to be a complete listing of all organizations which sell products, nor are they an endorsement of actual products.

Chapter 9
Remedies

Introduction

All of the information that has been discussed up to this point can be considered impractical unless it can be put into practice. Formulating a comprehensive ergonomic program is complicated and must take into account a variety of factors. Within the context of discussing the office environment, we are dealing specifically with two complementary issues: (1) how the vision of the worker can be maximized to produce the most positive effects in the workplace; and (2) how the workplace can be arranged to enhance the visual abilities of the worker. At this point we will review the various areas of the workplace and discuss the possible remedies for problems which may exist. Please keep in mind that each workplace has its own unique setup so this discussion will be generalized for a variety of situations.

Reflections and glare

A study of eye care professionals (Sheedy, 1992) found that the most common complaint of VDT-users was of glare. Glare is created by improper lighting in the workplace. The two main sources are from light directly shining into the eye (direct glare) or from reflections from surrounding surfaces (reflected glare).

Direct, or discomfort, glare primarily comes from bright sources of light in the field of view. Light often leaves an overhead fixture in a wide angle, resulting in light directly entering the workers' eyes if they

Figure 9.1 Glare from overhead lighting in the field of view can be very annoying and affect performance.

are viewing their work in a near-horizontal viewing angle. This problem can be resolved, at least partially by lowering the height of the monitor, so that the direction of gaze is more toward the desk rather than the surrounding room.

To test for direct glare, a simple shielding test should be employed. Have the worker viewing in their normal working position. Then simply ask them to shield their eyes with their open hand, as if simulating the bill of a cap. They will notice an immediate difference in the comfort level of their eyes; possibly even the physical relaxation of the muscles around the eye. If the source of light is to one side, the hand should be placed on that side of the eyes. If this test reveals a significant subjective sense of 'relief' from the worker, then the light source should be redirected or shaded.

Reflected glare is another significant factor in worker discomfort. This comes from various surfaces in the workplace, including the display screen itself. Some of these sources include white paper on a desk, white desktop surfaces, desktop lamps which are improperly positioned or even brightly-colored clothing which can be reflected in the screen. To discover if there is glare directly on the display screen (called specular glare), have the worker sit in their normal position with the screen off. Any bright surfaces or lights will be reflected from the screen and become visible. Adjusting the offending source, using muted colors, or repositioning the monitor will solve this problem.

Anti-glare filters which fit over the display screen are readily available. The purpose of the anti-glare filter is to decrease the luminance of the blacks, thereby increasing contrast. A Cornell University study (Hedge, 1996) found that 80% of anti-glare filter users reported that filters made it easier to read their screens and more than half said the filters helped their productivity. It must be

Figure 9.2 Shielding the eyes from offending light sources can allow your eyes to relax.

pointed out, however, that anti-glare filters themselves are not necessarily the entire solution to visual stress on VDTs. One study of over 25,000 employees (Scullica *et al.*, 1995) reported that glare filters alone do not reduce the occurrence of symptoms consequent to various factors, such as refractive defects, time spent at a VDT and monitors' characteristics.

In general, there are two types of filters: glass or plastic and mesh. The mesh filters are made from a black cloth material that is tightly drawn over a frame which fits over the VDT screen. This acts somewhat like a screen door – allowing direct light to pass through while blocking out oblique rays, like those which may come from scattered light. Mesh filters can be very effective at making the screen blacker and they are usually less expensive than glass filters. However, they also have a major drawback. The worker must now be sure to view the screen in exactly the correct position. If the tilt of the monitor does not allow for a parallel screen face, the image will be severely compromised. Additionally, the worker is viewing an image which has a degraded resolution. In general, the gains of improved contrast outweigh the loss of reduced resolution resulting in a net improvement but this compromise does not occur with the glass filters.

Glass or plastic filters primarily act as neutral density filters, the general transmission level of which is about 30%. Since reflected light must pass through the filter twice, the luminance of the reflections is reduced more than tenfold (30% × 30% = 9%). This significantly improves the contrast yet, since the worker is viewing through a clear optical element, there is no significant loss in the resolution of the screen characters. Some glass or plastic screens have a polarizing property which can provide further improvement, but only if there are specular reflections (as described above) in the screen.

Glass and plastic filters must be coated with an anti-reflection coating so that they do not create more reflections. Without a coating, a glass surface will reflect incoming light, which will appear as an additional glare point. It is critical for these screens to be kept clean because any smudges or dirt reduce its anti-reflective property. When using any anti-reflective screen, the display brightness is reduced. This necessitates increasing the screen brightness. It should be adjusted so that it is similar in brightness to the immediate visual surroundings. Some monitors may not have enough range of adjustment to compensate for the effects of the filter.

The American Optometric Association (AOA) has established a program to provide evaluation and recognition of quality VDT glare reduction filters. Products that meet the minimum specification established by the AOA Commission on Ophthalmic Standards are allowed to display the AOA Seal of Acceptance in product labeling and marketing. Specifications cover construction quality, image quality, glare reduction, reflectance and the ability to withstand environmental testing. The listing of approved companies and models which have been approved by the AOA can be found in Appendix E.

Workplace conditions

There are as many workplace settings as there are different facilities for each business. It would be impossible to address every possible workplace condition and make accurate recommendations in this brief discussion. However, there are some general guidelines which every business should address to enhance the working environment and vision of the office and industrial worker.

Lighting

With regard to the visual system, the most significant factor in workplace performance is lighting. As we discussed in Chapter 3, lighting is too often overlooked and misunderstood in many office settings. Unfortunately, many office designers are either unaware of how an office is to be used or the plans change between the initial design stage and the final layout. Controlling the lighting of the workspace is critical in maintaining proper visual efficiency. There are a number of steps that can be taken to ensure that lighting control is available and used properly. Many of these proposals are inexpensive and easy to implement.

1. Turn off the offending lights. If you notice glare on a screen or from a light source, turn it off. Most offices have been designed for paper use and are overly lit for VDT use. Keep the ambient light to about 40 to 50 foot-candles, as opposed to the 80 or 100 recommended for general office use. One should also be aware of 'hot spots', which are often sources of glare. Diffuse or indirect lighting creates a softer and more uniform visual field.

2. Many offices have two wall switches controlling the fluorescent light fixtures. Each switch turns on half of the bulbs in all of the fixtures. Most commonly, both switches are turned on for the day. Turning one switch off, thereby cutting illumination by one-half will, at first, elicit a negative reaction from the employees. However, with time, possibly just a few minutes, they will notice that there is less tension around the eye area and will appreciate the new level.

3. Fluorescent light fixtures can be retrofit with parabolic louvers which direct the light straight down into the room rather than scatter it in various directions. The louver results in the light being directed in a single direction, hopefully where it is needed and not into the eyes of another worker in a different part of the room.

4. Re-orient the work station so that bright lights are not in the field of view. Sometimes the work desk can be rotated 90–180° so that the fluorescent lights or bright windows are not in the field of view. Many workers make the mistake of placing their VDT right in front of a window, wishing to capitalize on a scenic view. This is a situation in which the brightness differences between the screen and the window can be a thousand times greater than those which cause discomfort glare. In a large office with rows of overhead fluorescent lights, it is usually better if workers view along the row of lights instead of viewing across the rows. This can easily be tested by observing the brightness, or measuring the luminance, in each orientation. This is because more light usually leaves the light fixtures laterally. Therefore, the fixtures will appear less bright if you are viewing along their length.

5. Shield your eyes. Often, if the ambient lighting cannot be altered significantly, using a visor (as described above) can shield the eyes adequately to provide comfort. A worker can wear a visor for a day or two as a test to determine if the discomfort is really a result of the glare from an offending light source.

6. Avoid bright reflective surfaces. In some work environments, the desk tops are white. This should be avoided because it results in an additional source of glare. Desktops and other furnishings should have a matt finish. Ceilings should be painted white and walls should be medium light.

7. Use blinds or drapes on windows. This can be rather controversial because the worker usually wants to have a view to the outside if it is available. However, if the outside is considerably brighter than the objects in the room (as it usually is), then the window is serving as an additional source of glare. The best solution is usually blinds, the most preferable being verticals because they can be adjusted to allow for a view while re-directing glare.

8. Use task lights. Care must be taken to use a task light which has the correct properties so that it does not create additional glare. If the light is scattered and falls on the VDT screen, then it will wash out the image and make matters worse. To select a good task light consider whether you need a symmetrical or asymmetrical distribution of light. The close work of architects and designers is best done with symmetrical light; VDT users work best with asymmetrical, thus enabling the worker to direct the light to the paper copy but not at the display screen.

One of the better task lights is the *PL* compact fluorescent lamp which is offered in many sizes and wattage with high color rendition, good energy saving and long life. Some people prefer halogen lights for their pristine light output, compactness and excellent beam control. A word of caution, however: halogen creates an intense heat that needs proper shielding and is both expensive and fragile. The better task lights also have flexible arms and various mounting options with which you can direct the light.

9. Use partitions. Very often the offending light sources can be eliminated from the field of view by erecting or moving partitions.

10. Check the brightness of the screen. The background of the screen should match, as evenly as possible, the immediate

background illumination. This is usually accomplished quite easily with a display with black characters on a white background. It is not possible to accomplish this if the display has light characters on a dark background. This setting is not recommended because of the extreme difference in illumination between the screen and the surroundings.

Ambient workplace lighting is also a very important consideration. With so many offices lit by fluorescent tubes, the potential for flicker is a concern. Flicker is common when the power source is weakened or the ballast or tube is too old. However, it is also possible that productivity and efficiency can be affected by a normal, fully charged fluorescent light. One study (Wilkins *et al.*, 1989) tested normal fluorescent lighting (50Hz lamps that produced 100Hz flicker) and high frequency ballast lighting (32kHz) on a group of office workers. Even though there were no perceptible differences in the two lighting conditions, eyestrain and headaches occurred significantly less frequently with the 32kHz lighting than with the 100Hz lighting. This indicates that a flicker that is beyond our perceptual abilities can contribute to headaches and eyestrain.

As a general rule of thumb, lights should be changed when they have reached about two-thirds of their life expectancy. When the lights are installed, the life expectation can be calculated by taking the average number of hours per day (and night) that the light will be on multiplied by the number of days to equal the life expectancy of the lamp (listed on its packaging). Multiply this number by two-thirds and mark that date on the calendar so the lights can be replaced regularly.

Viewing height and distance

The placement of the monitor and hard copy has been a source of intense debate among those who have an interest in ergonomics. The early standards had the display screen positioned 18 to 28 inches away at eye level – often achieved by setting the monitor on top of the CPU of the computer and just behind the keyboard. Eye care professionals, however, quickly realized that this is not a normal viewing posture for the visual system and began investigation into what viewing posture will reduce eyestrain symptoms.

The distance for viewing the VDT is the first factor which should be considered. While our normal book/paper reading distance is approximately 16 inches, the VDT requires a different distance largely due to its size and illumination. The ANSI/HFES-100 standards recommend a minimum viewing distance of 30cm (12in). However, this is not the *preferred* distance but just the *minimum*. A closer look at the standard (see Chapter 12) reveals that optimum viewing distance for VDT users is best determined following an evaluation of several factors. If you sit at 12 inches from your screen, you will realize that this 'feels' too close. The screen size was also a factor which was not considered in this standard. Larger screen sizes dictate different recommendations regarding viewing distance. In general, the larger the screen, the larger the font you can view, the further back you can sit.

This brings up the discussion of font size. A font is a style of printing and the size of that text can be varied greatly within a computer system. There are formulas for determining the maximum viewing acuity of the eye at any particular distance but this calculation is

Remedies

beyond the scope of this book. However, there is one general rule which should be kept in mind. Called the '3X Rule', it means that the minimum height of text should be three times the 'threshold' of visual acuity at that point. For example, if you have 20/20 vision at 25 inches, then the best size target would be three times that, or 20/60. In terms of font sizes, that equates to about a 10 point type at 24 inches. A practical way to calculate this is to have your screen type visible and move back as far as you can before it starts to become unreadable. Measure that distance and then move to one-third of that distance. For example, if the maximum measured distance is six feet, then sitting two feet (24 inches) away is the proper distance. These figures can vary depending on monitor quality and refresh rates but serve as a good general rule to follow. The best rule, however, is if you can't see the letters, make them bigger (don't move closer).

When employees performed paper tasks at a work desk, there was little concern for the viewing distance and placement of the paper. However, with the VDT being set in an upright position on the desk, viewing height has also undergone scrutiny in recent research. The optimal height of the screen depends upon several factors, primarily how a person adapts to the vertical location of their task by changing their gaze angle (the elevation of the eyes) and/or by changing the extension/flexion of the neck. The details of the research in this area have been discussed previously (Chapter 3) but the recommendations are be discussed here.

It appears that small amounts of neck flexion and ocular depression are preferred. These effects would dictate that the height of the screen should maintain a lowered level. However, an additional element is the angle of the upper torso relative to gravity. For example, if the torso is angled backward, such as when leaning back in a chair, then this would be subtractive to neck flexion and ocular depression in terms of its effect upon screen height relative to the eyes. It becomes more apparent that there can be little separation between the posture of the body and the posture of the eyes. The research overwhelmingly indicates that the position of the monitor should be located below the horizontal plane of the eyes. It is probably best that it be located so that it is 10–20° below the eyes. This allows an appropriate amount of eye depression; neck flexion is compensated for by torso tilt and this amount is consistent with preferences established by research findings. At a typical distance of 24 inches, the screen should be located so that the *center* (assuming that the center of the screen is viewed most frequently) is 4–9 inches lower than the horizontal plane of the eyes.

With the advent of larger monitors, many of these 'hard and fast' rules must be re-evaluated. If a worker uses a 21-inch monitor at 24 inches, it would be rather difficult to maintain the center of the screen low enough without having to make significant alterations in the desk and chair heights. Once again it becomes apparent that the entire work space needs to be considered as a whole unit, rather than trying to piece many little ideas together. These larger screens also present another complication: screen tilt.

Screen and copy tilt

Simply lowering a display screen to its 'proper' position, though possibly an easy arrangement to achieve, is not the only position to consider. Screen tilt is also an important issue. Brand and Judd (1993) had subjects perform an editing task on a computer screen which was angled away at the top by 12 degrees from a perpendicular to the desk surface. Unfortunately, this study did not specify the location of the eyes with respect to the screen, so it is not known what the angle of the screen was relative to the line of sight of the subject. However, performance times and subjective preferences were better for the reference document being similarly angled (12 degrees) compared to conditions of 30 and 90 degree (flat on the table) tilt. Their study also demonstrated that the preferred angle of tilt of the screen was such that it is parallel with the frontal plane of the face.

As just mentioned, the larger monitors present another problem concerning screen tilt. These monitors are so deep in design that there is often not enough elevation between the base and the edge of the screen. The design of the workstation must be altered to allow for a higher chair position or lower desk level which, of course, presents other concerns. Ergonomic desks which have flexibility of adjustments are advantageous in these situations.

Because the work space is now taken up with a large piece of electronic equipment, document placement becomes a concern. In general, whatever is viewed most often during daily work should be placed straight in front of the worker. This applies to the display screen and/or the reference documents – whatever is viewed most frequently. Although this seems self-evident, many VDT workers situate their work so they are constantly looking off to one side.

The location of the reference material can be very important. Many VDT workers place the display screen straight in front of them and then locate the reference documents flat on the table next to the screen. This is not very efficient. This requires large and frequent eye, head and upper torso movements to look back and forth from the reference documents to the screen. A good solution is to purchase a document holder on a spring load, which can locate the reference documents adjacent to the display screen. It is also best to locate the reference documents so that they are at the same viewing distance from the eyes as the screen. In this way a focusing change is not required to look from one to the other. An alternative design is to locate the reference document so that it is between the monitor and the keyboard. This arrangement allows the eyes to focus and converge in their normal plane of motion, i.e. 'down and in'.

Ergonomic furniture

Even though this discussion is centered around the eyes and visual system, it should be evident by now that it is nearly impossible to separate the eyes from the rest of the body. Upper body position and eye position are intricately linked and one depends upon the other. In order to maintain a proper body position which maximizes productivity and reduces stress, proper desks, chairs and accessories should be considered.

The following suggestions assume independently adjustable work surface and keyboard height and an adjustable chair. They are simplified for a typical case and should always be modified based on job and personal factors. Also, they are for furniture adjustments only and for VDT work only.

1. Start with feet flat on the floor or on a foot rest.

2. Adjust chair height to a comfortable position that keeps the feet on the floor, thighs approximately parallel to floor. Forward sitters may prefer to sit higher than this, recliners may prefer to sit lower. People in knee-tilt chairs tend to prefer higher positions than people with column-tilt chairs that lift the feet when reclining.
3. Move all the way back in the chair and adjust the backrest height to a comfortable position.
4. Move the backrest in or out and adjust the angle so that the depth of the seat is appropriate. The user should be able to reach the backrest without experiencing any pressure behind the knees at the front of the seat cushion. Large users should not feel that the seat depth is short and unstable.
5. Re-adjust the backrest height if necessary.
6. Adjust the chair tilt tension so the user can recline mostly by a weight shift rather than pushing off with the feet. If the heels rise significantly off the floor when reclining, the user is pushing with the feet too much and the tension should be lightened. If the user prefers not to recline, the tension can be kept tight or, in some chairs, locked.
7. Adjust the height of the armrests. They should be high enough so the elbows are supported without having to slouch or lean to one side. They should be low enough to avoid uncomfortable pressure under the forearms or elevation of the shoulders.
8. Adjust the width of the armrests. They should be close enough to avoid 'spraying' of the elbows as much as possible. If maintenance-adjustable, they should be far enough apart to allow the user to get into and out of the chair easily.
9. Adjust the keyboard height and angle so the user is comfortable and the wrists are in a neutral (straight) position. This does not necessarily mean having the keyboard at elbow height as long as the keyboard's angle reflects the angle of the forearms. In other words, a very low keyboard position can be used if the keyboard is tilted back. If the keyboard cannot be lowered appropriately, the chair may have to be raised and a footrest added but this should be a last resort.
10. Place the mouse where it can be used without reaching, preferably next to the keyboard. If using a keyboard tray, an extended

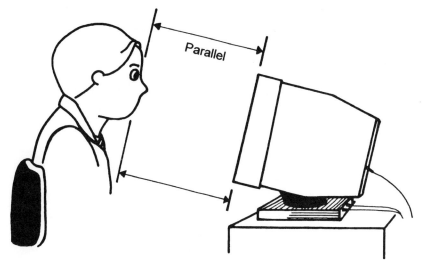

Figure 9.3 The angle of the monitor should approximate the angle of your face: they should be almost parallel.

Brightness, contrast and color

top may be necessary. The height of the mouse should be such that the wrist is fairly straight when using it.

The advantage of using a video display to view text and graphics is that there is flexibility in adjusting the image for maximum viewing. However, it also means that there is much room for error and degrading of the image. Enhancing the image of what is being viewed is a critical task in ensuring comfortable viewing which is free of eyestrain. The three general areas to consider are the brightness, contrast and color of the image.

These areas of concern have been previously discussed in detail (Chapter 4) so this should serve as a review of factors to consider. When viewing a dark background screen, be sure that there is relatively dim surrounding ambient lighting. This assures that the eye will equally accommodate to the illumination of both the display screen and the room in general. The brightness of the letters should be significant – approximately 10 times that of the background – but not bright enough to cause the edges of the letters to blur. The color of the letters, in a dark background display, should be white.

It is common for the luminance of the lighting fixture to be 100 times greater than that of the display that the worker is viewing. The Illuminating Engineering Society (IES) has established guidelines for the luminance of workplace lighting in the form of ratios between differing work surfaces. The ratio of the task to the immediate surroundings (approximately 25 degrees) should be 3:1 and that of the task to the more remote surroundings should be 10:1. (IES, 1981). Table 9.1 illustrates how the luminance of various surfaces compare with the VDT task.

While viewing a white background display, the contrast of the letters should be maximized by using black letters. The brightness of the background will more closely approximate that of a normal working office, usually having 75–100 foot candles of illumination. The goal here, again, is to have the background of the display screen approximate the illumination of the ambient lighting of the room. Even if using a color monitor, it is best to maintain a black on white condition if performing word processing duties. There is no 'optimum' color for use on a display screen. However, one must keep in mind

Table 9.1 How luminances of various surfaces compare with the VDT display

Object	Luminance (cd/m^2)
Dark background display	20–25
Light background display	80–120
Reference material (approx. 75 fc)	200
Reference material w/task light	400
Blue sky (window)	2,500
Concrete in sunshine	6000–12,000
Fluorescent lights (unshielded)	1000–5000
Auxiliary lamp (direct)	1500–10,000

that color vision distortions can occur if using a consistent color for long periods of time.

The best advice is to purchase a monitor which has simple and accessible controls for the brightness and contrast. If working with natural daylight (not usually recommended), the lighting will change throughout the day and the brightness of the screen should be changed accordingly.

Work habits

The three main factors which create visual stress in a VDT viewing situation are: (1) the visual condition of the worker; (2) the work environment and (3) the work habits of the worker. This last condition is probably the one factor most under control of the VDT user. For some reason, which has so far yet to be determined, viewing a video screen attracts the attention of the viewer more so than any paper object. This attention causes a change in the way we view the display screen – a change which is detrimental to our visual system. This 'stare' condition can be counteracted by using a '3-B Approach': Blink, Breathe and Break.

Because of the attention drawn to the screen, we 'forget' to blink. Unfortunately (in this instance), blinking is an automatic reflex and beyond our conscious efforts. A study (Tsubota and Nakamori, 1993) showed that there is less blinking while viewing a display screen than other tasks. This study was discussed in detail in Chapter 5 but it appears that one of the more logical reasons for decreased blinking during VDT work has to do with the height of the monitor. Because it is in a higher field of the viewing area, the eyelids maintain a larger distance between them. This creates a larger gap between the lids which makes the blink reflex more of an effort. Lowering the screen will reduce the size of this gap and allow the lids to blink easier. Blinking 15–20 times a minute is normal for most situations; 10–15 times a minute is fine for VDT viewing.

Breathing is also another reflex but is subject to our psychological and emotional states. We often hold our breath when in stressful situations. One classic example is that of a weight-lifter. When attempting to lift the weight, he or she must hold their breath to build up tension in the muscles and increase their strength. Our bodies perform this type of action on a regular basis, holding back the breath whenever we attempt some stressful feat or encounter stressful situations. Keeping the breath flowing in an easy, smooth movement allows the muscles to relax while still being used efficiently.

When office procedures still involved typewriters, pencils and paper, there was a great deal of physical activity in the work area. Hands were moving in all directions to insert paper, return the typewriter carriage, grab a pencil, or turn a page. Office workers were getting up to make copies, deliver papers to another office, look for carbon paper and other types of physical tasks. With the integration of computers into the workplace, much of these movements are accomplished by the push of a button. We hear from ergonomic experts that we should be working in a 'neutral' position. However, they forget to tell us to get out of that neutral position regularly. Our bodies are designed for movement and should be moved routinely.

This holds especially true for the visual system. Our eyes have many muscles associated with them. If these muscles are 'stuck' in the same position for an extended period of time, they will adversely affect vision. Taking visual breaks is a very easy thing to do because they do not involve leaving the desk and do not have to be long in duration. A *micro* break consists of simply looking at a distant object every 10–15 minutes or so. This should be done for only 15 seconds; not enough to adversely affect the work being performed. A *mini* break should be performed about every 30 minutes and consist of closing the eyes or performing a simple eye exercise (see the last section of this chapter). This should only last a minute or two. A *maxi* break should follow along with your routine work breaks where you get up and move around for an extended period of time. This should be taken at least every two hours and last for at least fifteen minutes. This scheme will allow the eyes to change their viewing condition regularly and still allow the worker to produce effectively.

Computer glasses

Most prescribed eyeglasses are for general purposes: driving, movies, TV, shopping, allowing the wearer to perform a variety of tasks. However, there are also *task-specific* lenses which are made to allow the wearer to do a specific task. Computer glasses are designed with these types of lenses. However, there is more than one type of lens available and it may actually not be a complicated lens to wear.

A 'computer' prescription is any lens which allows the wearer to see the display screen clearly and comfortably. They do not, however, necessarily allow clear vision at any other distance. There are a number of combination lenses which may allow for this type of vision. The following discussion will look at some of the more popular task-specific lenses available and their appropriate applications and limitations.

Single vision

A single vision lens is any lens with one focal point. The only difference in any single vision lens is its application for the particular visual condition of the wearer. For example, a person who wears glasses all day long to see clearer at a far distance is using a single vision lens. Likewise, 'reading' glasses are also single vision lenses, appropriate for the reading distance. So, for the particular application of seeing a display screen clearly, a single vision lens can be designed so that the clear viewing distance is the same as the display screen. All lenses have a range of clear focus: the stronger the prescription of the lens, the shorter the range of clear focal power. It is desirable that the display screen (or any point of near viewing) falls in the middle of that range. For this reason, it is imperative that the prescribing doctor be aware of the viewing distance for the screen. The drawback of these lenses is that, although the screen may be clear, most other viewing distances will be out of focus when viewed through the lenses.

Intermediate/near bifocals

The normal bifocal lens is designed so that the distance viewing portion is on the upper part of the lens and the reading portion is on the lower part. In this instance, the wearer simply lowers their eyes to read. However, the lens can be modified so that the upper part of the

Remedies

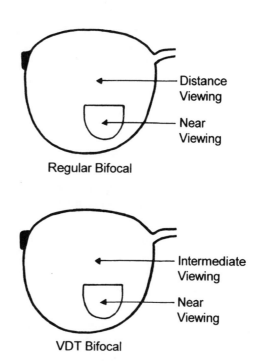

Figure 9.4 Typical and task-specific types of bifocals.

lens is prescribed with the display screen power and the lower part of the lens with the reading power. This design can be effective but is somewhat limited in the available range. The lower part may not be adequate for keyboard viewing if needed.

CRT trifocals

In the late 1970s, when computers where becoming more commonplace, a lens was developed which was supposed to cover all the needs for VDT workers. Taken from the concept of an *executive* bifocal (which covered the entire width of the lens), this computer lens was a *tri*focal which offered three separate focal distances: distance, intermediate and near. The optical properties of the lens were not very desirable but it did serve the purpose of having three distinct focal distances. However, proper fitting was often an issue and the very obvious lines between the segments were not very attractive. The lenses were also rather heavy to wear.

Progressive addition lenses

Developed in the 1950s, and enhanced in the 1970s, was a lens which had no distinct line between the various power ranges. This was often

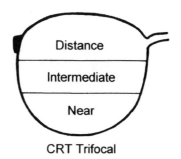

Figure 9.5 Early CRT trifocal lens with three viewing distances.

incorrectly referred to as the 'no-line bifocal' because of the lack of any lines on the lens. However, this is more than just a bifocal because the lens maintains a number of different focal powers to clear an object at a full range of vision, from distance to intermediate to near. This lens design has exploded onto the marketplace and there are currently over 25 manufacturers making about 60 varieties of progressive addition lenses.

The standard wear progressive lens has its distance viewing area on the top part of the lens and gradually increases the power as the level of gaze is lowered, so that reading at a near point is possible while viewing through the lower part of the lens. It has proven to be a viable alternative to traditional bifocals and trifocals. There are some adaptation concerns (there are some peripheral distortions) but once a wearer adapts to the lens the viewing is clear and comfortable. With the increased popularity of computer use, especially by the presbyopic (over 40) population, lens manufacturers have attempted to adapt the progressive lens to computer use. There are some valid designs currently available, each of which has its own positive aspects.

The Technica® by American Optical Company can be considered a traditional progressive lens design which has been 'displaced' to accentuate the intermediate portion of the lens. The lower part of the lens allows for clear near-point viewing, the middle and upper part of the lens are for intermediate viewing. There is a very small segment at the very top of the lens which allows for distant viewing. This distant portion is very small and not adequate for general use, like walking or driving. It can be useful for viewing a clock across the room or recognizing someone walk into a room. The lens has experienced some degree of popularity and can be worn successfully. There is, however a level of distortion in the lower peripheral areas which necessitates the wearer to still do some degree of 'nose pointing' when viewing objects off to the side. Depending on the setup of the workstation, this lens can be very effective.

The Readables® by Varilux Corporation has a slightly different design. It is fitted so that there is a gradual *increase* of power below the midpoint of the lens to allow for clear near viewing; and a *gradual* decrease of power above the midpoint to allow for a clear intermediate distance viewing. On the surface this sounds like a valid concept. However, the drawback is that there is no single clear point in the lens – the power is *always* changing. Thus, the slightest head movement can throw it out of focus for a particular distance. This lens has no distance viewing area.

Most recently, Sola Optical Company released a newly designed lens called Access™. This lens has two general viewing areas: the near and intermediate. Between them is a small, gradual change with no lines. Both viewing areas cover the entire width of the lens, giving the greatest coverage in lateral viewing. Although this lens design is relatively new it appears to have great potential in fitting the needs of a wide variety of display screen viewers.

Computer user frames

In addition to lenses for specific tasks, there is a frame which has been developed for various near viewing tasks. Called *Focal Change*, it is a standard ophthalmic frame which has a nose bridge which is readily adjustable so it can raise or lower at a slight touch. This has the advantage of relocating the bifocal of the lens into a different reading position. One task which would benefit from this alteration is newspaper reading, which often involves viewing a close reading distance at an elevated level of gaze. Whether or not it is appropriate to VDT viewing is still debatable since the power of the bifocal portion of the lens would need to be altered to accommodate both the normal 16-inch reading distance and the 20–30 inches VDT viewing distance. This would be difficult to achieve and it seems that the frames would have a limited usefulness in VDT situations.

Lens tints

When computers were first introduced into the workplace, vision professionals assessed the screen attributes and, combined with the subjective complaints of glare and eyestrain, assumed that a tint in the computer glasses would be appropriate for VDT users. At that time, there were three general VDT screen designs being used: green letters, amber letters or white letters, all on a black background. Using various theories of light behavior and optical physics, it was postulated that there should be one of three different tints used for viewing the VDT depending on which screen arrangement was being used. A person using green letters was prescribed a light magenta tint; amber letters dictated a blue tint, while white letters required a neutral gray tint. It is not necessary to discuss the concept behind these recommendations because their theories have proven to be inaccurate.

The only lens color that might be required in a general workplace is the light rose (or pink) colored tint. The reason behind this recommendation is that it tends to cut the glare and mute the green colors which predominate in a workplace using standard fluorescent lighting.

Anti-reflective coatings

A normal eyeglass lens only allows 92% of light to pass completely through it. Four percent is reflected by the front surface of the lens while an additional 4% is reflected by the back surface. However, a coating can be applied to the front surface of the lens which effectively cancels out this reflection and allows 99% of the light to pass through.

This has a double benefit to the general eyeglass wearer. First, it allows the wearer's eyes to be more visible by someone looking at them. This condition is similar to that which you might experience when viewing an eyeglass wearer on television and seeing the reflections of studio lights on his or her lenses so that their eyes are not visible at all. Secondly, the anti-reflective coating, by allowing more light to pass through a lens, also makes the view through them more distinct. This is most noticeable by the wearer viewing lights at night, whereby they might normally see glare around street lamps or oncoming headlights. The anti-reflective coating allows the lights to pass through without the extra glare or distortion.

In the VDT viewing environment, the anti-reflective coating will allow the viewer to see the screen more clearly. However, it will not affect (or might worsen) the glare from other environmental lighting. The anti-reflective coating tends to be more difficult to maintain because it is subject to smudges in normal daily use. If they can be kept clean and unscratched, lenses with anti-reflective coatings make all types of viewing easy on the eyes.

Eye exercises

In Chapter 2, we discussed the general anatomy and physiology of the eyes and visual system. It should be apparent now that the eyes are an integral part of the central nervous system. Although accurate eyesight is developed shortly after birth, there is a developmental aspect to vision – it changes and develops as we do. Because of this intimate relationship, the visual system is subject to change and alteration in relation to its environment.

The system of eye exercises is more accurately termed *vision therapy*. Vision therapy is an integrated program of techniques and procedures that assist the person in improving all aspects of vision, including general coordination, balance, hand-eye coordination, eye movements, eye teaming, form recognition, visual memory and imagery. The program is more than 'exercises', a term which seems to imply the strengthening of muscles. True vision therapy is an individualized program of progressively arranged conditions of learning for the development of a more efficient and effective visual system. It can be used to improve a number of eye conditions.

For the VDT user, vision therapy can improve the skills necessary to view the display screen with comfort and efficiency. The program can be combined with a prescription for VDT viewing glasses which can assist the eyes in achieving the proper balance. Some of the skills which are required by VDT users are:

Tracking: the ability to follow a moving object, such as a mouse pointer, moving from one section of a page to another, smoothly and accurately with both eyes.

Fixation: the ability to quickly and accurately locate and inspect with both eyes a series of stationary objects, one after another, such as moving from word to word while reading.

Accommodation: the ability to look quickly from far to near and vice versa without momentary blur, such as looking from the display screen to the clock on the wall or someone walking into the room, and back to the screen.

Peripheral vision: the ability to monitor and interpret what is happening around you while you are attending to a specific task with your central vision; the ability to use visual information perceived from a large area.

Binocular coordination: the ability to use both eyes together – smoothly, equally, simultaneously and accurately; includes the ability to converge the eyes, which means aiming them toward each other when looking at near objects, and to relax the convergence, which means to move the eyes away from each other as the eyes re-focus from near to distant objects.

Hand-eye coordination: the ability to use your hands and eyes together in a synchronized manner so that a task like following a mouse pointer can be performed with efficiency.

Maintaining attention: the ability to keep doing any particular skill or activity with ease and without interfering with the performance of other skills.

Near-vision acuity: the ability to clearly see, inspect, identify and interpret objects at near distances (within 20 feet of your eyes).

Remedies

Visualization: the ability to form mental images in your 'mind's eye' and retain them for future recall or for synthesis into new mental images beyond your current or past experiences.

Relaxation: the ability to relax the eyes and the visual system; important for preventing and treating eyestrain.

Here are some techniques to attempt. It's usually best to do the techniques early in the day, before your eyes are too tired. Don't try to do them all every day. Try all of them over the course of a few days and see which ones are the most difficult for you; then concentrate on those for a period of a few weeks and see what improvements you can make in your vision.

Accommodative rock

The accommodative rock exercise helps improve the eyes' ability to change focus and see clearly near and at a distance. Accommodation is the process by which the eye changes focus and is probably the most important and most often performed function of the eyes. The ability to focus decreases with age, but adequate focusing ability can be maintained for longer periods of time with exercises like this one. To do the accommodative rock:

1. Put some large letters (such as a large newspaper headline) on a wall and stand back 20 feet (use glasses if necessary to see these letters).
2. Take some small-print letters (such as from a newspaper article) and hold them in one hand. Cover one eye with the other hand (don't close it).
3. Bring the small print as close as you can while still being able to see it clearly. Stop at that point and move the letters back about one-half inch.
4. Look at the large letters on the wall again. Are they clear?
5. Look again at the small print (keep it at the same distance). Clear again? This change in focus (accommodation) should only take a second.
6. Switch back and forth for a few minutes until you can clear the distance and near letters easily. Try it with each eye. When this becomes easy, move the small letters one inch closer and repeat the procedure.
7. Do this exercise for five minutes with each eye twice a day, preferably getting in both sessions before evening tiredness sets in. You can also try this exercise in many different situations throughout your day, whenever you find yourself with a near and distant (over 20 feet away) object on which to focus; for example, try switching back and forth between your wristwatch and a wall clock.

Rotations

Rotations increase the eyes' tracking ability and help your ability to pay attention to an activity. Smooth eye movements are basic to good vision. To do this activity, you'll need a marble and a pie tin.

1. Put the marble in the pie tin and tilt the tin so that the marble can roll around the edge of it.
2. Hold the pie tin with the marble about 16 inches from your eyes and roll the marble around the edge of the tin at a steady pace.
3. Follow the marble with your eyes only, without moving your head.
4. See how well you can concentrate on this exercise for several

minutes at a time. Have someone watch your eyes to see how smoothly they are moving.

5. Do this exercise once a day; do it in one direction for two minutes and then in the other direction for two minutes (you'll get dizzy if you keep the marble going in the same direction for more than a few minutes).

Alphabet fixations

Alphabet fixations improve your ability to enter the eyes on an object in an instant – to fixate on it. This is the type of eye movement used in reading. This exercise also helps with near-vision acuity. To practice alphabet fixations:

1. Cut out two strips of paper. Type or clearly print the alphabet in a vertical column on each strip.
2. Hold the strips about 18 inches away from you and a bit farther apart than shoulder width.
3. Call out the letters of the alphabet, alternating from one strip to the other ('a' from one strip, then 'a' from the other; 'b' from one strip, then 'b' from the other and so on). Keep your head pointed straight between the strips and absolutely still as you do this.
4. Start to spell words by using the letters from alternating strips. For example, spell the word 'boy' by taking a 'b' from one strip, an 'o' from the other strip and a 'y' from the first strip. Be sure you see the letters before you call them out! Your speed should increase with practice.
5. Do this exercise once a day for five minutes.

Monocular (one-eyed) fixations

Monocular fixation exercises enhance the ability to fixate using one eye at a time. This particular activity is also designed to work on hand-eye coordination. You'll need a string, a small ring (such as one you wear on your finger or a key ring) and a knitting needle or long pencil for this exercise.

1. Attach the ring to the string and hang it at eye level (from a doorway, for example).
2. Stand about two feet away and cover your left eye.
3. Step forward with your right foot and, using your right hand, try to put the pencil or knitting needle through the ring without touching it. Try this several times.
4. Cover your right eye, step forward on your left foot and, using your left hand, try to put the pencil or needle through the ring.
5. Master this exercise using a stationary ring; then give the ring a push and try going through it in the same manner, only while it's swinging. This is also great practice for many sports.
6. Do this exercise for five minutes once a day.

Wall fixations

The wall fixations exercise improves your ability to fixate and your peripheral vision at the same time. You'll need eight white index cards, three inches by five inches each, a felt-tipped marker, a blank wall and maybe some gentle music.

1. Draw one two-inch high number in the center of each index card, using a felt-tipped marker and a bold stroke, so that you have eight index cards, each with a number from one to eight.
2. Fasten the cards to a blank wall in an eight-foot square. Vary the

Remedies

spacing of the numbering of the cards so that your eyes have to jump past some cards to see the next number.

3. Stand about six feet away from the wall and directly in the center of the card pattern. Put a book on your head to keep your head steady, and cover one eye.

4. Starting at the card with the '1,' begin shifting only your eyes to each card in numerical order. You may want to have gentle music playing to help keep your eye movements smooth. Keep your head very still and look directly at each number, being aware of the other numbers in your peripheral vision.

5. Move gradually closer to the wall as you improve at this skill. As you do this, your eye movements will have to become more extreme.

6. Switch to the other eye and repeat this procedure.

7. Try this exercise using both eyes together.

8. Do this exercise once a day; do it for two minutes with each eye alone and then two minutes with both eyes.

Marsden ball

This is a great activity for improving eye tracking and fixation as well as hand-eye coordination and attention maintenance. You'll need a red rubber ball about four inches in diameter, some strong thread or string and a ball-point pen.

1. Write letters randomly all over the ball with a ball-point pen.

2. Pierce the ball with the thread or string and suspend it so that the ball can swing freely.

3. Cover one eye, give the ball a slight push and try to touch one letter at a time as you call it out. Keep your head as still as possible as you do this.

4. Switch to the other eye.

5. Do this exercise once a day; take two minutes for each eye.

Peripheral awareness

Developing your peripheral vision awareness will allow you to expand your vision beyond your central concentrated task. This helps to stimulate various parts of the retina and can help to keep you from staring at one point. It is important to maintain awareness of the other parts of your visual field. See Figure 9.6 for this exercise.

1. Hold the target at your normal reading distance.

2. Maintain steady fixation on the center point of the card.

3. While looking only at the center point, observe and record the letters which you can detect peripherally. Be sure your central vision is still on the center point.

4. Keep trying to expand your peripheral awareness to include more letters. Peripheral vision is not as sharp as central vision so don't be discouraged if this seems like a difficult procedure.

Deep blink

The deep blink exercise is designed to improve your ability to accommodate (change your eyes' focus) and your distance acuity. It's also a relaxation technique. If you feel dizzy or faint at any time during this exercise, stop and rest. You'll need a wall and some large letters, such as the ones in large newspaper headlines.

1. Fasten the large letters to the wall.

2. Stand a few feet back from the letters on the wall and, *without* your glasses or contacts, gradually move back until the letters start to blur. This is the distance at which you'll start working.

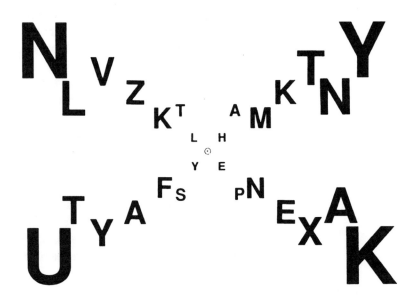

Figure 9.6 Look directly at the center of this figure. Use your peripheral vision to view each series of letters as far out to the side as you can.

3. Sit in a chair in a relaxed posture; take a deep breath and let it out slowly. Repeat this a few times until you feel relaxed.

4. Take a deep breath and hold it. With your breath held, close your eyes, clench your fists, and tighten the muscles in your whole body – legs, arms, stomach, chest, neck, face, head and eyes. Squeeze these muscles very tightly for about five seconds.

5. Snap open your hands and eyes after about five seconds, exhale quickly through your mouth, and relax your entire body. Breathe slowly and look at the letters, blinking gently as necessary. Stay very relaxed and try to look *through* rather than *at* the letters. After a second or two the letters should clear. (If you felt dizzy or faint after the clenching-and-relaxing part of this exercise, just do the relaxation part on the next try. Take slow, deep breaths and practice looking *through* the letters on the wall.)

6. Take a step backwards (if the letters remain clear) and repeat this procedure from the new position.

7. Keep going back farther and farther, one step at a time, and see how far back you can go while still keeping the letters clear (no glasses or contacts!). You may be amazed to find that after a few weeks you can stand quite a few feet further back than where you started and still see those letters.

8. Do this exercise at least once a day.

Brock String exercise

The Brock String exercise helps develop good depth perception and improves peripheral vision and binocular coordination. You'll need a piece of string four feet in length.

1. Tie a knot in the middle of the string.
2. Attach one end of the string to any object that is at your eye level (you can sit or stand for this activity).
3. Hold the string between your thumb and forefinger, stretch it taut, and hold it up against your nose.

Remedies

4. Look at the far end of the string. You should see an 'A' without the crossbar. You should see the knot in the middle of the string as two knots, one on each side of the 'A'.

5. Look at the knot in the center of the string. It may take a few seconds, but you should be able to see an 'X' pattern with one knot in the middle.

6. Shift your gaze back and forth from the 'A' to the 'X' pattern and back again until it is smooth and requires little effort.

7. Move your gaze up the string toward your nose when you can shift back and forth with the 'A' and 'X' patterns. As you do this, you'll find the center of the 'X' moves up toward your nose too. When you get very close to your nose, the 'X' becomes a 'V', and the center knot should now appear to be two knots on the two sides of the 'V' in your peripheral vision.

8. Practise shifting from 'A' to 'X' to 'V' until you can feel your shift flowing smoothly along the string. As you improve, shorten the string, but keep the knot centered.

9. Do this exercise once a day for five minutes.

Convergence stimulation

Convergence is the aiming of the eyes toward each other as you look at near objects. This skill is critical for all near-point activities, such as reading. The convergence stimulation exercise helps develop binocular coordination (of which convergence is one aspect) and good depth perception. This exercise also helps your ability to focus your attention. You'll need Figure 9.8 and a pencil.

1. Focus your eyes (with corrective lenses if necessary) on the illustration at your normal reading distance. You should see two sets of concentric circles.

2. Lay the point of your pencil between the two sets of concentric circles and focus on it, staying at your normal reading distance.

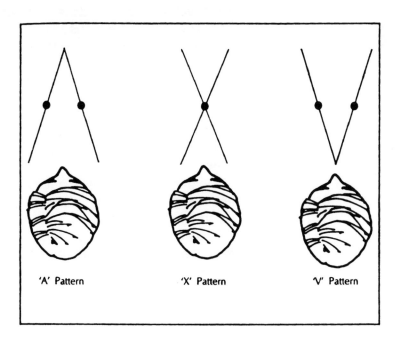

Figure 9.7 The Brock String should appear as an 'A' when viewing the distant point; as an 'X' when viewing the mid-point; and as a 'V' when viewing the close point of the string.

3. Move the pencil slowly toward your eyes, leaving the illustration where it is. Keep focusing on the pencil, but be aware of the circles beyond the pencil as you do so.

4. You should begin to see three, rather than two, sets of concentric circles when the pencil is approximately six inches from your eyes. The set in the middle (the one that's not really there) should appear as a three-dimensional figure; the smaller circle should appear to be farther away from you than the larger one, as if it were a cup or flowerpot. Keep the pencil still and keep looking at it. Your eyes should feel like they're crossing (they're actually just pulling in toward each other).

5. When you can accomplish seeing a middle set of circles with a three-dimensional effect, relax your focus and look away from the circles for a second. Then try the exercise again and see if you can regain the image. This may take some practice.

6. Do this exercise until it is easy to maintain and hold the center image. Then practise it a few times each day, alternating it with the convergence relaxation exercise (see next exercise).

Convergence relaxation

Convergence relaxation is the opposite of convergence stimulation. It's the ability of the eyes to relax (actually to diverge from their converged position) as they focus on distant objects. Excessive near work can cause the eyes to have trouble relaxing, so this is a good exercise to break up your day if you do a lot of reading or computer work. The convergence relaxation exercise helps with binocular coordination (the convergence relaxation aspect of it), depth perception and attention maintenance. You'll need Figure 9.8 and a blank wall.

1. Focus your eyes at a blank wall at least 10 feet away from you.

2. Bring the illustration of the concentric circles slowly into your line of sight at your normal reading distance, but keep your eyes focused at the wall. (You can hold the illustration just above a point on the wall where your eyes are focused.)

3. You should begin to see three sets of concentric circles, as you did in the convergence stimulation exercise. But, this time, the three-dimensional effect on the center set of circles should have the circles looking like upside-down cups or flowerpots, that is, the smaller circle should appear to be closer to you than the larger circle. (You may see the other sets of circles this way too, in your peripheral vision.)

4. Relax your eyes for a second or two, when you have achieved this effect, and try it again.

5. Do this exercise until it becomes easy for you. Then practise it a few minutes each day, alternating it with the convergence stimulation exercise.

The convergence stimulation and convergence relaxation exercises should be done together because they are exercising the same set of eye muscles, but pulling them in different directions. If one of these exercises is much easier for you than the other, concentrate on the one that's harder until you can do it as easily as the other one.

Palming

Palming is a relaxing measure to use between other exercises and throughout the day. It allows you to relax your mind and your eyes

Figure 9.8 View the pencil tip with your central vision. These circles will double then 'fuse' when the eyes are converged sufficiently. At that point, they will appear in three-dimensional view.

because you're not focusing on anything but blackness. Palming can also increase your visualization skills.

1. Cover your closed eyes with your hands. Keep the palms over, but not touching, your eyelids. Your fingers should overlap near your hairline and there should be enough room to breathe easily. Rest your elbows on a table. Complete blackness should be all you see. If you see flashes of light, just let them go and allow the blackness to return. You can either continue to focus on the blackness, or you can now start to visualize a relaxing scene of your own choosing.

2. Take a deep breath and feel the muscles around your eyes completely relax.

3. Breathe deeply and slowly eight times and do this exercise at least eight times a day, preferably *before* you start intense near-point visual tasks, such as reading or computer work.

Head rolling

Head rolling is another relaxation technique which is a good general body exercise. It increases blood flow, which increases life-maintaining oxygen, to the brain and eyes – and it feels good.

1. Let your head gently fall forward while seated. Then, slowly roll it around from one shoulder to the other, making a complete circle. Keep your shoulders level and maintain regular, but deep breathing.

2. Change the direction of your head roll. Roll two or three times in each direction.

3. Try this first thing in the morning and again later in the afternoon for a relaxation break. This is also good for a short mini-break from your computer work day.

This and other general body exercises are discussed in the next chapter.

References

Brand, J.L. and Judd, K.W. (1993) Angle of hard copy and text-editing performance. *Human Factors*, 35 **(1)**, 57–69.

Hedge, A. (1996) The effects of using an optical glass glare filter on computer workers' visual health and performance. Independent Survey, sponsored by Softview Corp.

IES Lighting Handbook (1981) Illumination Engineering Society of North America, New York, NY.

Scullica, L., Rechichi, C. and DeMoja, C.A. (1995) Protective Filters in the Prevention of Asthenopia at a Video Display Terminal. *Percept Mot Skills*, **80**, 299–303.

Sheedy, J.E. (1992) Vision problems at Video Display Terminals: a survey of optometrists. *Journal of the American Optometric Association*, **63**, 687–692.

Tsubota, K. and Nakamori, K. (1993) Dry Eyes and Video Display Terminals. *New England Journal of Medicine*, **328**, 8.

Wilkins A.J., Nimmo-Smith, I., Slater, A.I. and Bedocs, L. (1989) Fluorescent Lighting, Headaches and Eyestrain. *Lighting Research and Technology*, **211**, 11–18.

Chapter 10

General eye care tips

Introduction

This chapter is intended to be a sort of 'catch-all' to bring together some peripheral ideas and concerns regarding computer use. We will examine such ideas as nutrition, aging, contact lenses, exercises and more. It may appear on the surface that some of these ideas have nothing to do with VDT use but these are areas that can affect your ability to use a VDT. These concepts can be considered an aspect of 'holistic' health care where it is advantageous to consider the whole being or process when attempting to heal the organism.

Stress and vision

Stress is something we live with every day. It has become an integral part of our work day, as well as other aspects of our lives. It has positive and negative effects on our biological system, many of which are beyond the scope of this discussion. However, we will touch on the subject to emphasize the importance of stress reduction in maintaining good vision and good working habits.

We normally associate stress with muscle tension. While the two are actually different processes, it has become commonplace to accept the notion that stress usually leads to increased (excessive) muscle tension. There are many areas of our bodies where this

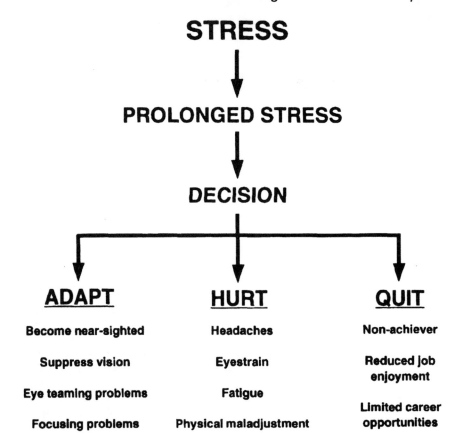

Figure 10.1 Our eyes will adapt to stress in a variety of ways (after Godnig, 1990).

muscle tension tends to manifest. Since this is a discussion of the visual system, the discussion will be limited to the visual stress associated with VDT work.

Stress can easily affect vision. Yet measuring how this occurs is not a routine clinical process since stress is often simply a subjective symptom. But since the eye and visual system are directly connected to the brain, measuring brain waves can be an indication of visual stress. The electroencephalogram (EEG) has been used to measure the brain waves of people experiencing visual stress (Pierce, 1966). It was found that there are changes in other bodily functions, including heartbeat, respiration, blood vessel size, as well as an overflow to other non-visual brain areas that can occur with a visually-demanding task. The earliest EEG studies in the 1920s confirmed that visual input affected brain wave patterns. Scientists have established not only that the EEG pattern could be changed by repetitive visual stimulation at a known frequency, but also that the brain would quickly respond by falling into that same frequency. This process is employed in practice during vision therapy to elicit a relaxation response by a patient. The frequency of stimulation is in the range of 8–12Hz, which approximates the alpha rhythm, or relaxed state of the brain. One might speculate what effects, negative or otherwise, might be elicited by the viewing of a 60–75Hz flashing image (the VDT screen) during the course of an eight-hour workday.

The reaction to visual stress can take many forms. One is to avoid doing the stressful task. Another is to do the work but approach it with

General eye care tips

reduced comprehension while experiencing physical discomfort. A third is to physically adapt to the stressful situation. This third adaptation could include the development of myopia or the suppression of one eye's image. Figure 10.1 shows the various adaptations to visual stress one might adapt.

Billette and Piche (1987) and Bergman (1980), among others, have noted a significant relationship between stress and job function. Often it is the job design and not the worker which is the cause of stress. Some jobs are inherently more stressful than others. For example, an air traffic controller, who is responsible for the safety of thousands of people on a daily basis, must be considered differently than a mail clerk who may occasionally glance at a display screen. In general, the more controlled, timed, repetitive and socially isolated the job, the more stressful it tends to be. Jobs which provide more latitude for the worker to control the job pace and design of the work are the easiest to deal with.

In addition to the organization of the work, other environmental factors affect the stress level of the worker. Factors leading to job stress include poor lighting, excessive noise levels, unhealthy air quality, inadequate work space, and poorly designed furniture and workstations.

Eye health concerns

VDTs, like all kinds of electrical devices, give off a small amount of electromagnetic radiation. This radiation is energy that moves through space at various frequencies. The radiation spectrum is divided between high-frequency ionizing radiation and lower-frequency non-ionizing radiation. X-radiation, such as that used for medical purposes, is a type of ionizing radiation while sunlight consists of several types of non-ionizing radiation including visible light, ultraviolet and infrared.

Several types of radiation may be produced by VDTs, including x-radiation, infrared (heat), visible light, radio frequency, ultraviolet and others. The cathode ray tube functions using a stream of electrons which energizes the phosphors to cause illumination. Side products

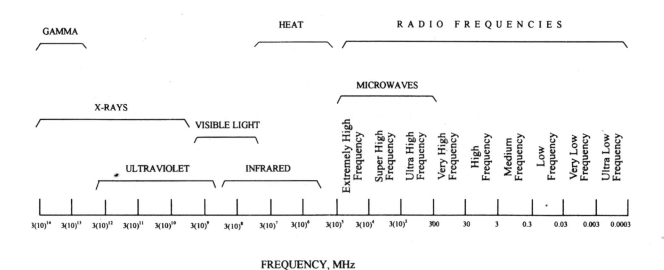

Figure 10.2 The electromagnetic spectrum.

of that process include several types of optical radiation including ultraviolet, visible and infrared radiation. The electrical circuitry, including the transformer, produces infrared and various types of radio frequency radiation.

It is very difficult to determine what subtle effects, if any, low-frequency fields may have on living tissue over long periods. It is known that the body's cells have their own electric fields, and some laboratory studies have shown that these internal fields can be disrupted by exposure to even low-energy electromagnetic fields (EMF). Some scientists hypothesize that subsequent cell changes – notably in cell membranes, genetic material, immune function, and/or hormonal and enzyme activity – may lead to increased risk of cancer. It is, however, difficult to extrapolate from 'test-tube' studies to human beings living in the real world.

The stream of electrons by which a VDT displays images on its screen generates fields in the very-low frequency (VLF) and extremely-low frequency (ELF) ranges, which pass right through the machine's case. They are the magnetic fields that scientists are most concerned about and that are hardest to shield. Various appliances produce similar energy fields, but the distinctive type from VDTs is emitted in sharp bursts which may have a greater effect on tissue.

Concern about VDT radiation began in the early 1970s when reports appeared about computer operators having high rates of headaches, miscarriages, cataracts and other health problems. There have been various reports of findings for and against caution but thus far no conclusive evidence has shown that there are any significant long-term effects from VDT use. The National Institutes of Health concluded that the apparent miscarriage rate for VDT users was not caused by any effects of the VDT itself. The reports of VDT users having a higher-than-normal incidence of cataracts have also been disputed as a side-effect of the many users who are elderly, when cataract formation is more common. No substantiated negative health effects on the eye has been shown to be caused specifically from VDT use.

On the side of caution, however, it is probably prudent to minimize the risk of exposure, especially if you are a full-time user. Electromagnetic radiation falls off rapidly with distance from the source. Try to maintain a 24–28-inch working distance from the screen and also be sure that there are no other monitors in your immediate vicinity. It has been shown that more radiation comes from the back or sides of the monitor than through the screen. There has been no evidence that warrants ultraviolet protection in glasses which may be worn for VDT use. It has been shown that the amount of ultraviolet light emanating from the screen is minimal and no cause for concern. When shopping for a monitor, make sure it complies with the 'Swedish standards', which most new monitors do. Also, do not succumb to anti-glare screen advertisements which claim that they block 'radiation'. Although these may block low-frequency electric fields, they do not block magnetic fields, which are a greater concern.

Aging factors

The Bureau of Labor Statistics estimates that there will be an increase of over 42% of workers who are 55 years old or older

General eye care tips

by the year 2004. It would be highly unlikely that these older workers will be doing a manual labor type of work. Instead, the trend for the older worker is to do some type of desk work which involves the use of a computer. Let's examine the various considerations which the older workers must address in their VDT viewing environment.

The most significant condition facing the vision of the older worker is presbyopia. As you recall from Chapter 2, presbyopia is the loss of accommodation with age, which usually becomes apparent after the age of 40. The presbyopic person must hold their reading material farther away if they hope to see it clearly. One advantage of using a VDT in the workplace is that it is most often (as it should be) placed farther away from the eye than is normal paper reading material. For the pre-presbyopic viewer who does not use reading glasses, this could delay the symptoms of presbyopia. However, once 'reading' glasses are prescribed to compensate for the loss of accommodation, caution must be taken to assure that the entire viewing area is available.

Since a reading distance of 16 inches is the standard testing range for eye care professionals, that is the area of clear near viewing obtained through reading glasses. If a VDT screen is set at 28 inches or so, it may not be legible through the glasses. If a presbyopic person uses bifocals, the reading portion is normally situated in the lower portion of the lens to allow for clear reading in a lowered-eye position. The display screen is most often situated in a higher visual field position, thereby requiring a bifocal wearer to tilt up their head if they wish to read through the bifocal lens to see the screen. This can (and usually does) create a problem with neck and shoulder pain.

Even the use of the progressive addition lens (PAL), often referred to as the 'no-line bifocal', is not usually acceptable for prolonged VDT viewing. The intermediate range segment of the lens is often too narrow to provide extended viewing comfort. Using computer glasses, with or without lines, which are designed specifically for the working distance and viewing angle of the VDT screen will usually resolve most problems.

Lighting in the workplace can also become a significant factor with the aging of the workforce. A 60-year-old worker needs ten times the amount of light as a 20-year-old. If the workplace has younger and older workers in the same area, a proper balance of light should be available to please both age groups. Task lighting usually provides the best resolution to this problem.

Dry eyes are a very common condition in the elderly population, especially for females. It appears that changes in hormone levels can adversely affect the tear layer formation, leading to a dry eye condition. While it is difficult to make specific recommendations on what to do for dry eyes, artificial tears and vitamins are but two remedies which have shown some success. The artificial tear is a temporary measure but is relatively successful in reducing the symptoms of dry eyes. The vitamin option is discussed in a subsequent section.

Computers and contacts

We briefly touched on the subject of contact lenses while discussing the industrial work setting (Chapter 7). In addition, there are some considerations which should be noted for the VDT worker who wears contact lenses in the course of their work day.

It has already been pointed out that dry eye is one of the significant symptoms of Computer Vision Syndrome. This condition can be exacerbated greatly by the use of contact lenses. Since the contact lens is a piece of plastic floating on the tear layer of the eye, it can be susceptible to dry environmental conditions. If the lens wearer has a marginally dry eye condition, wears contact lenses and uses a VDT regularly, then a dry eye feeling is likely to manifest during or after VDT use. The regular use of an artificial tear substitute eye drop is recommended to relieve this symptom.

Often the room in which the VDT is used is an arid environment due to the requirement of the CPU of the computer. The CPU must not be exposed to a high humidity environment, so the air in these rooms can be especially dry. This, again, can lead to noticeable symptoms of dry eyes while using contact lenses. The VDT screen itself has an electromagnetic charge and therefore attracts dust to it. These dust particles can often lodge under a contact lens causing discomfort. Lens removal and re-wetting is a simple solution to this problem.

For visual correction, contact lenses are routinely prescribed to correct distance vision problems. Wiggins and Daum (1992) found that the incomplete correction of astigmatism in contact lens wearers using VDT screens created symptoms of visual stress. If this distance prescription is adequate for near/intermediate use as well, no problems should be noted. However, if the near/intermediate vision correction is different than that for distance, the contact lens wearer may experience other symptoms of Computer Vision Syndrome. It is possible that computer glasses might be prescribed to be worn over the contact lenses.

Vitamins and VDTs

This section is not intended to offer specific remedies so that your VDT looks better or lives longer by giving it a specific vitamin regimen. However, it is intended to offer *you* some suggestions that you might want to incorporate into your own diet which might relieve some symptoms which may be experienced while viewing your display screen.

Let's take another look at the dry eye condition. There are two causes for dry eyes: (1) not enough tears (quantity); or (2) rapidly evaporating tears (quality). It is relatively rare for a dry eye condition to be caused by a decreased quantity of tears. Many successful contact lens patients have a very low volume of tears. The normal tear volume is maintained by the glands within the lining of the eye. But the tears you experience while crying are produced by a gland in the upper corner of the bony orbit just behind your eyebrow. The quality of the tears – which maintains their integrity so they do not evaporate too quickly – is the more important concern. This tear quality is dependent on a fatty layer which covers the watery layer of tears. If this fatty layer is deficient, it will break up quickly, allowing the water tear layer to evaporate too quickly, leading to a dry eye symptom. One of the most important factors in maintaining this fatty layer is Vitamin A (or beta carotene). Patel *et al.* (1991) showed that Vitamin A is essential in maintaining the stability of the tear film on the eye.

General eye care tips

Vitamin A is just one of the group of nutrients known as antioxidants. This group of vitamins and minerals is useful in reducing the effect of oxygen free-radicals on living tissue. The foundation of this research is beyond the scope of this book but there are some key issues which should be kept in mind. The group of antioxidants include Vitamin A (beta carotene), Vitamin C, Vitamin E, selenium, inositol, panthothene and zinc. These have shown to be effective in reducing the severity of cataract formation and the condition of age-related macular degeneration. More research is being conducted as of this writing but it looks very promising that vitamin and mineral supplements can assist our resistance to certain types of diseases and disorders.

General body exercises

Throughout this text we have maintained the notion that the eyes are an integral part of the body and must be treated in much the same way – with attention and care. Since the 'eyes lead the body', it only makes sense that good body exercises will help you to maintain good posture at the workstation. Here are some general body exercises which are easy to do and very effective. A word of caution, however: if you have any pre-existing condition which might be aggravated by doing these exercises, please consult with your physician prior to attempting them.

A. **Pectoral stretch** Do this when you find yourself slouching. Clasp your hands behind your head. Tuck in your chin, press the back of your head into your hands and push your elbows as far back as you can. Hold for 3 seconds, then relax and repeat 5 times.

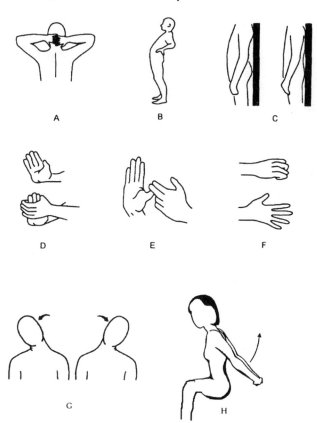

Figure 10.3 *Try these easy exercises to get your blood flowing properly and keep your energy while using your VDT.*

B. **Disc reliever** Do this to reverse effects of repetitive or sustained bending. Place your hands in the hollow of your back. While focusing your eyes straight ahead, bend backwards over your hands without bending your knees, then immediately straighten up.

C. **Pelvic tilt** Do this to reverse effects of standing with 'sway back'. Begin by standing with your back to the wall. Tighten your stomach muscles to flatten your back. Hold for several seconds. Once you've mastered the exercise, do it sitting or standing.

D. **Wrist/finger** Hold one hand with fingers upward. Gently push fingers and wrist back with the other hand. Hold for 3 seconds. Repeat 5 times for each hand.

E. **Thumb** Hold one hand with fingers upward. Gently pull back the thumb with the fingers of the other hand. Hold for 3 seconds. Repeat 5 times for each hand.

F. **Whole hand** Spread the fingers of both hands apart and back while keeping your wrists straight. Hold for 3 seconds. Repeat this exercise 5 times for each hand.

G. **Head roll** Relax your shoulders and pull your head forward as far as it will go. Hold for just 2 seconds. Then slowly rotate your head along your shoulders until it is all the way back. Continue rolling around to the other side until you return to your original position. Roll your head in one direction 3 cycles, then reverse the direction for another 3 cycles. Feel the upper shoulder muscles relax. Do these slowly and feel the stretch in the neck muscles.

H. **Shoulder squeeze** Another excellent stretch for slouchers. Lace your fingers behind your back with the palms facing in. Slowly raise and straighten your arms. Hold for 5–10 seconds. Repeat 5–10 times.

While doing all of these exercises, it is important to remember to maintain a full and smooth breathing pace. Full breaths allow for further relaxation of the muscles being stretched. They also allow for increased blood circulation which will improve your alertness and mental activity.

References

Bergman, T. (1980) Health effects of video display terminals. *Occup Health & Safety* (Nov/Dec), 24–28, 53–55.

Billette, A. and Piche, J. (1987) Health problems of data entry clerks and related job stressors. *J. of Occupational Medicine*, **29** (12), 942–948.

Godnig, E.G. and Hacunda, J.S. (1990) *Computers and Visual Stress*. Seacoast Information Services, Charlestown: RI.

Patel, S., Asfar, A.J. and Nabili, S. (1991) Effect of visual display unit use on blink rate and tear stability. *Optometry & Vision Science*, **68** (11), 888–892.

Pierce, J.R. (1966–67) Research on the relationship between near point lenses, human performance, and physiological activity of the body, Optometric Extension Program Courses, Research Reports and Special Articles, **39** (1–12).

Wiggins, N.P. and Daum, K.M. (1992) Effects of residual astigmatism in contact lens wear on visual discomfort in VDT use. *Jour American Optom Assn*, **63** (3), 177–181.

Chapter 11

The economics of visual ergonomics

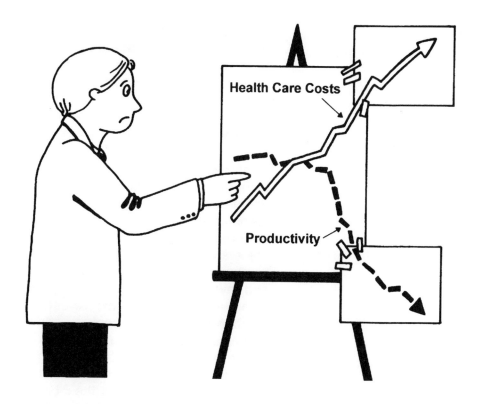

Introduction

Just like every aspect of corporate management, ergonomics must be a fiscally sound aspect of the business. The 'bottom line' for every business is making a profit and if decision makers are to spend a significant amount of money on a program, there must be an equally significant return on that investment. Ergonomics, however, is a relatively new concept for many corporations and they may not be aware of how it can improve their profitability.

Even more obscure is the concept of *visual* ergonomics – the interaction of vision with the workplace. It is not an easy concept to promote to corporate executives who often want to take little responsibility for employee well-being. Yet it is a real concern which can affect their bottom line. Let's take a look at the obvious, as well as subtle, ways that a solid ergonomic program – general and visual – can be a valuable asset to any company.

Cost justification for ergonomics

The first thing which must be acknowledged is that ergonomics is about money – costs and savings. It would be wonderful to think that just the value of having a stress-free working environment would be enough to motivate companies, but that just isn't reality. The driving force behind ergonomic programs is this: what do they cost and what do they save? A good place to start is with statistics about the real costs involved with CTDs in the workplace. According to a Liberty Mutual Insurance Company study, the average cost of a case of upper extremity CTD is $8,070. The average cost of an occupational back injury is $8,321. If an employee requires surgery, those figures can quickly escalate to 10–15 times those amounts. The Bureau of Labor Statistics (BLS, 1994, Internet) reported that repetitive motion injuries were the leading cause of lost work days in 1994. According to the survey, 36% of employees suffering repetitive motion injuries lost at least 31 days of work. The overall average loss was 18 days. Companies do understand that lost days equals lost money. OSHA studied companies implementing ergonomics programs and showed that injuries can be eliminated for about 60% of employees.

Whether called carpal tunnel syndrome, repetitive stress injury or cumulative trauma disorder, wrist, hand, elbow, shoulder and back injuries continue to be an almost unavoidable plague for workers using computers in manufacturing, service and office businesses.

Increasingly, companies are focusing on proactive ergonomic techniques to prevent or contain such injuries in their early stages and reduce costs. Even in this era of corporate downsizing, human resource and loss prevention professionals are able to obtain money for ergonomics training and equipment from skeptical finance officers by explaining the potential for fewer, less costly workers' compensation claims, increased productivity, healthier employees and fewer lost work days. OSHA has targeted cumulative trauma disorders and 'electronic sweatshops' in proposed ergonomic standards. The standards are complete but implementation had been stalled because business groups have labeled them onerous. These are discussed in the following chapter.

Businesses may be reluctant to have standards imposed on them, but many have improved their bottom line by making ergonomic improvements on their own. For example, the U.S. Postal Service estimates workers' compensation claims from carpal tunnel syndrome alone cost between $18 million and $29.5 million from 1992 through 1995; costs for individual cases ranged from $13,000 to $33,000. The Postal Service developed and tested an ergonomics program in its automated mail sorting sites in 1996 that has resulted in savings of more than $10 million, according to a report on the program.

The West Bend Mutual Insurance Co. in West Bend, Wisconsin saw productivity among more than 400 employees increase by 16% when it moved into a new, ergonomically correct building in 1991. Since then, carpal tunnel complaints have dropped from about 14 a year in the early 1990s to a single case in 1995.

Pitney Bowes Inc., a $3.6 billion multinational company based in Stamford, Conn., saw an average 33% decrease in workers' compensation claims for cumulative trauma disorders in seven divisions between 1994 and 1995. The company has mail sorting, copier, financial and management services businesses. Claims decreased by 54% in one division and by 58% in another.

Initiating a general ergonomics program because of safety and health issues is fine for safety and health professionals. However, the corporate management must see it from a corporate perspective, which often revolves around other factors. If your company has a safety and health program in place, then including ergonomics is a simple matter of alignment of purpose. If the values of a good health and safety program are not currently employed, the long-term prospect of reducing CTDs and RMIs is unlikely to become a priority.

Historically, ergonomics has been an engineering discipline with a slant on human performance. At one time, ergonomics was about developing appropriate work methods, supportive work environments and unique hardware design to reduce stress and promote productivity. Yet, over the past decade or so, coincidentally with the onset of increased VDT use, ergonomics became more concerned with preventing CTDs. Many of the more subtle and deep-rooted concerns were cast aside in favor of the easy fix of a wrist guard.

To begin to incorporate a solid ergonomic program into the corporate environment is to start with a more palatable description. It has been suggested that 'high-performance workplace' be used to indicate that performance (or bottom line) is the target of the program. It is better to address 'the role of an effective work environment in increasing productivity' rather than describe why 'the current work environment is hurting people'.

It is almost ironic how much money has been invested to improve employee health: employee fitness centers, nutritionally-balanced menus, non-smoking work areas, stress management, and so on. Yet it is interesting to note that the workplace itself has received very little attention. One can make the analogy of someone learning all the important aspects of competitive swimming: the proper stroke, the best breathing techniques, the proper conditioning, but if there is not enough water in the pool or the water is too cold, then all the training in the world won't make a difference. The workplace is the fundamental structure for employee performance. The order of remediation should be: (1) fix the environment; (2) protect the person; and (3) medical management of the problem.

Companies tend to record the number of CTDs and accidents for OSHA reporting. This method of 'managing' CTDs is a retroactive approach which lends itself to a downward spiraling cycle. People struggle with their work environment but have some good days and some bad days, until they are unable to perform and report an illness or problem. The difference between what people are capable of doing and what their physical work environment allows them to do is the productivity gap. That is the real cost of poor quality. It limits the performance of the company by millions of dollars.

Executives worry about corporate productivity and corporate productivity is about human performance. Top companies realize this and integrate employee education and skills development into the proposals for the upcoming year. The goal of ergonomics management is to improve productivity and to improve the workplace as a part of the company business plan. Improving the work environment is a business function.

Providing eye care to employees

It should be apparent that vision problems are a very real and significant concern for employees using display screens. A study by NIOSH showed that 22% of VDT workers have musculoskeletal disorders while the symptoms of Computer Vision Syndrome are apparent for approximately 75–90% of VDT workers. The causes of these visual symptoms are a combination of visual problems and poor office ergonomics. An individual may have a marginal vision problem which does not present a problem while doing less visually demanding tasks. However, when this person performs a visually intensive task, the symptoms are often manifest.

The eye and vision problems can largely be resolved through management of the visual environment and by providing proper eye care for employees. But why should an employer invest resources to resolve these problems? The primary reason is that it is just good business. Business executives are familiar with investing money in processes or equipment which improve efficiency. Although we typically think of assembly lines and blue collar workers when we talk about work production, we must recognize that people sitting before a display screen are a major part of work production today. It is important that business improve the efficiency of the office worker today, just as assembly line processes were streamlined in the past.

Let's take a look at the economics of this issue. Since working at a computer is a visually intensive task, and the sense of vision is used to acquire the information needed for job performance, it is reasonable to expect that improvements in the display or in the visual capabilities of the user will work towards improving performance efficiency. There are several studies which show that better displays or better vision result in improved efficiency. VGA displays – the most common format of DOS-compatible equipment – has a pixel density of approximately 75 dots per inch (DPI). It has been shown that increasing the pixel density on the screen from 75 DPI to 115 DPI results in 17.4% faster reading performance for 30-minute reading sessions (Sheedy, 1992). Also, reading speed improvements of 4.1% to 19.9% (depending on display type) have been shown for adding a gray scale improvement of image quality (Sheedy and McCarthy, 1994). This argues for providing better monitors, but it also argues for providing better vision of the worker. Certainly if subjects with good vision can obtain reading speed improvement with a better quality image, a person with poor vision will attain better performance by improving their vision.

Uncorrected vision problems in the work force create worse vision than those in the situation above that showed 4–19% decreases in visual task performance. Although they were laboratory studies, and the tasks were performed for durations that are considerably shorter than a full work day, it is likely that similar inefficiencies occur daily for workers with uncorrected vision disorders. We might even expect that 8-hour productivity would be more greatly reduced because of the symptoms and fatigue which accompany the vision problems.

If an employee's compensation is $30,000 (including benefits), a one per cent improvement in work efficiency is worth $300. Eye care can be provided for considerably less than this – and results in much *more* than a one per cent increase in productivity.

The costs of providing eye care to computer users can be calculated from the following data (Sheedy, 1996):

- 12.43% of optometric patients are examined primarily because of problems at a VDT (1991);

- 40.3% of those patients obtain spectacles specifically for work at the computer;
- there were 86 million eye examinations given in 1994;
- the average eye examination fee in 1994 was $54.00;
- the average retail price of glasses in 1994 was $133.51.

These assumptions result in an estimated calculated cost of $1.15 billion for providing eye examination and computer-specific glasses to VDT users. Lost work days for visual problems are seldom figured in statistical analysis and therefore are not a significant cost item. Decreased productivity due to uncorrected visual problems, however, is not included in this calculation. It appears that there are considerably more direct costs involved with musculoskeletal disorders ($5.8 billion) compared to visual problems ($1.15 billion), thus another reason why CTD disorders receive the majority of the attention.

Many companies already offer vision care to their employees as a benefit of employment. Even if a company offers vision care to all employees, it may not necessarily be meeting the needs of their VDT employees. Proper care of the VDT employee requires more than just a simple glasses check up, as we discussed in Chapter 6. Many VDT employees require a different pair of glasses for their work than that needed for their other daily visual needs. The employee is reluctant to use their employee benefits for a pair of computer glasses; they feel that the benefits should be used to provide for their general glasses.

In the interest of work efficiency, everyone who needs a visual correction should wear one. The best way to ensure this is for the employer to provide the eye wear for all VDT employees. However, many employers feel that the employees should be responsible for providing their own general eye wear, and that it should be the employers' responsibility only if the glasses are different from the general eye wear or if they would not otherwise be required. This can be accomplished, with cost savings, by establishing a list of diagnostic/treatment conditions (see Table 11.1) for which glasses will be provided. In order for glasses to be provided under the program, panel doctors would need to arrive at one of the listed diagnoses and determine that the glasses are different in prescription or design than those required for other daily visual needs.

Costs can also be controlled by establishing limitations on the frames and spectacle lenses which are provided. Single vision and bifocal lenses are necessary program options. General progressive

Table 11.1 Determining which visual conditions warrant special attention can assist in providing necessary care without excessive expenditures

Diagnostic Conditions Which Warrant Special VDT Glasses
Presbyopia
Accommodative Disorders
Hyperopia
Binocular Vision Disorders

lenses ('no-line' bifocals) should not be provided because they generally do not function well for VDT workers. Trifocal and specially designed progressive addition lenses can be very useful for many VDT workers. While it is desirable to provide these lens options, most users' visual needs can be properly managed with single vision or bifocal lenses, thereby resulting in cost savings. Tints and coatings, as we discussed previously, do little to help in solving the problems that VDT users have and are not necessary for the program. If only the basic lens options are to be covered, then employees should be able to pay the difference if they want more expensive options.

Another important cost control element is to provide good ergonomic assessment and correction where indicated. It is clear that many of the eye and vision problems which computer users experience can be resolved by evaluating and improving the visual work environment. Visual ergonomic evaluation and correction will help to reduce the utilization of eye care services.

Vision screening, as described in Chapter 6, can help to reduce over utilization by identifying those employees who are most likely to benefit from an eye examination. Professionally managed vision screening is costly and it is questionable whether the savings in utilization overcome the costs of performing a screening. Self analysis tools are available which more cost-effectively enable employees to screen themselves for vision problems and also to educate them about proper use of their eyes and their computer environment.

References

Bureau of Labor Statistics (1994) Characteristics of Injuries and Illnesses Resulting in Absences from Work. BLS: Internet.

Sheedy, J.E. (1992) Reading performance and visual comfort on a high resolution monitor compared to a VGA monitor. *Journal of Electronic Imaging* **1(4)**, 405–410.

Sheedy, J.E. (1996) The bottom line on fixing computer-related vision and eye problems. *Journal of American Optometric Association*, Guest Editorial **67:9**, 512–517.

Sheedy, J.E. and McCarthy, M. (1994) Reading performance and visual comfort with scale to gray compared with black and white scanned print. *Displays* **15(1)**, 27–30.

Chapter 12

Ergonomic standards

Introduction

Like most aspects of business, ergonomics is defined by legislation. However, since it is a relatively new area – at least for office ergonomics – the laws governing ergonomic requirements remain in a state of flux. Even the definition of ergonomics itself has undergone recent revisions and many lay people have never even heard of the word. Yet there is much happening in legislative areas which will affect the ergonomic community. This discussion will encompass some of the relevant laws pertaining to ergonomics as well as the standards on which many of them have been developed.

The Americans with Disabilities Act

Signed into law on 26 July 1990, the Americans with Disabilities Act (ADA) is a wide-ranging legislation intended to make American society more accessible to people with disabilities. It is divided into five titles:

Title I Employment Business must provide reasonable accommodations to protect the rights of individuals with disabilities in all aspects of employment. Possible changes may include restructuring jobs, altering the layout of workstations, or modifying equipment. Employment aspects may include the application process, hiring, wages, benefits, and all other aspects of employment. Medical examinations are highly regulated.

Title II Public Services Public services, which include state and local government instrumentalities, the National Railroad Passenger Corporation, and other commuter authorities, cannot deny services to people with disabilities participation in programs or activities which are available to people without disabilities. In addition, public transportation systems, such as public transit buses, must be accessible to individuals with disabilities.

Title III Public Accommodations All new construction and modifications must be accessible to individuals with disabilities. For existing facilities, barriers to services must be removed if readily achievable. Public accommodations include facilities such as restaurants, hotels, grocery stores, retail stores, etc., as well as privately-owned transportation systems.

Title IV Telecommunications Telecommunications companies offering telephone service to the general public must have a telephone relay service to individuals who use telecommunication devices for the deaf (TTYs) or similar devices.

Title V Miscellaneous Includes a provision prohibiting either (a) coercing or threatening or (b) retaliating against the disabled or those attempting to aid people with disabilities in asserting their rights under the ADA.

The ADA prohibits employment discrimination against 'qualified individuals with disabilities'. A qualified individual with a disability is an individual with a disability who meets the skill, experience, education and other job-related requirements of a position held or desired, and who, with or without reasonable accommodation, can perform the essential functions of a job. The ADA's protection applies primarily, but not exclusively, to 'disabled' individuals. An individual is disabled if he or she meets at least any one of the following tests:

1. He or she is substantially impaired with respect to a major life activity.
2. He or she has a record of such an impairment.
3. He or she is regarded as having such an impairment.

Other individuals who are protected in certain circumstances include (1) those, such as parents, who have an association with an individual known to have a disability, and (2) those who are coerced or subjected to retaliation for assisting people with disabilities in asserting their rights under the ADA. While the employment provisions of the ADA apply to employers of 15 employees or more, its public accommodations provisions apply to all sizes of business, regardless of number of employees. State and local governments are covered regardless of size.

An impairment substantially interferes with the accomplishment of a major life activity when the individual's important life activities are restricted as to the conditions, manner, or duration under which they can be performed in comparison to most people. For example, a person with a minor vision impairment, such as 20/40 vision, does not have a substantial impairment of the major life activity of seeing. It is not necessary to consider if a person is substantially limited in the major life activity of 'working' if the person is substantially limited in any other major life activity. For example: if a person is substantially limited in seeing, hearing or walking, there is no need to consider whether the person is also substantially limited in working.

In general, a person will not be considered to be substantially limited in working if s/he is substantially limited in performing only a particular job for one employer, or unable to perform a very specialized job in a particular field. For example: a person who cannot qualify as a commercial airline pilot because of a minor vision impairment, but who could qualify as a co-pilot or a pilot for a courier service, would not be considered substantially limited in working just because he could not perform a particular job. Similarly, a baseball pitcher who develops a bad elbow and can no longer pitch would not be substantially limited in working because he could no longer perform the specialized job of pitching in baseball. But a person need not be totally unable to work in order to be considered substantially limited in working. The person must be significantly restricted in the ability to perform either a class of jobs or a broad range of jobs in various classes, compared to an average person with similar training, skills and abilities.

The regulations provide factors to help determine whether a person is substantially limited in working. These include:

- the type of job from which the individual has been disqualified because of the impairment;
- the geographical area in which the person may reasonably expect to find a job;
- the number and types of jobs using similar training, knowledge, skill or abilities from which the individual is disqualified within the geographical area, and/or
- the number and types of other jobs in the area that do not involve similar training, knowledge, skill or abilities from which the individual also is disqualified because of the impairment.

For example: a computer programmer develops a vision impairment that does not substantially limit her ability to see, but because of poor contrast is unable to distinguish print on computer screens. Her impairment prevents her from working as a computer operator, programmer, instructor or systems analyst. She is substantially limited in working, because her impairment prevents her from working in the class of jobs requiring use of a computer.

The concept of 'reasonable accommodation' is a critical component of the ADA's assurance of non-discrimination. Reasonable accommodation is any change in the work environment or in the way things are usually done that results in equal employment opportunity for an individual with a disability. An employer must make a reasonable accommodation to the known physical or mental limitation of a qualified applicant or employee with a disability unless it can show that the accommodation would cause an undue hardship on the operation of its business. Some examples of reasonable accommodation include:

- making existing facilities used by employees readily accessible to, and usable by, an individual with a disability;
- job restructuring;
- modifying work schedules;

- reassignment to a vacant position;
- acquiring or modifying equipment or devices;
- adjusting or modifying examination, training materials or policies;
- providing qualified readers or interpreters.

However, an employer is not required to provide an accommodation that is primarily for personal use. Reasonable accommodation applies to modifications that specifically assist an individual in performing the duties of a particular job. Equipment or devices that assist a person in daily activities, on and off the job, are considered personal items that an employer is not required to provide. However, in some cases, equipment that otherwise would be considered 'personal' may be required as an accommodation if it is specifically designed or required to meet job-related rather than personal needs. An example of this is an employer who would not have to supply a personal item like eyeglasses. However, the employer might be required to provide a person who has a visual impairment with glasses that are specifically needed in order to use a computer monitor.

Most of the ADA concerned with the visual aspects of disabilities relate to low vision or visual impairment. This topic has been covered extensively in Chapter 8 and should be reviewed to determine what appropriate measures might be taken to accommodate the visually-impaired worker. Typical visual disorders, such as hyperopia, myopia, astigmatism, binocular or accommodative problems, do not qualify as visual impairments under this category because they are generally considered 'correctable with standard appliances or techniques'.

Technical standards

Ergonomic considerations are based on current technical standards and are governed by some form of legislation. However, since office ergonomics is such a relatively new area of concern, much of what is used is speculative or just not based on a standard scientific foundation. This section will discuss several technical standards which are used in developing ergonomic rules and we will attempt to sort out what is fact and what is fiction. The development of technical standards is based on the latest available scientific study and is continually open to revision. Therefore, some of this information may be updated by the time of publication.

Because this book is mainly concerned with visual ergonomics, this discussion will be limited to the visual, or ophthalmic, standards which are concerned with this area of study. These ophthalmic standards have been established for various reasons. They provide a standard nomenclature, definitions and terminology so that various segments of an industry can work together more effectively. Secondly, they provide quality control to assure safety to the public interest.

The American National Standards Institute (ANSI) Z80.1 standards for ophthalmic lens tolerances establishes the lens quality criteria for optical laboratories, optometrists and others. These standards are applied to ophthalmic lenses, eyeglass frames, contact lenses (both hard and soft), and ophthalmic instruments. ANSI Z87.1 is the standard for occupational and industrial eye and face protection that specifies the properties of eye protective materials. The impact resistance portion of ANSI Z80.1 specifies the minimum protective

Ergonomic standards

Table 12.1 Categories for ANSI standards concerning vision and eye wear

Standard	Description
Z80.1	Prescription Ophthalmic Lenses
Z80.2	First-Quality Rigid Contact Lenses
Z80.3	Nonprescription Sunglasses and Fashion Eyewear
Z80.4	Contact Lens Accessory Solutions for Conventional Contacts
Z80.5	Dress Ophthalmic Frames
Z80.6	Physiochemical Properties of Contact Lenses
Z80.7	Intraocoular Lenses
Z80.8	First Quality Soft Contact Lenses
Z80.9	Low Vision Aids
Z87.1	Occupational and Education Eye and Face Protection
Z136.1	For the Safe Use of Lasers
ANSI/ASTM	Specification for Eye and Face Protective Equipment for Hockey Players
F513-85 ANSI/ASTM	Specification for Eye Protectors for Use in Racquet Sports
F803-85 ANSI/ASTM	Specification for Ski Goggles
ANSI/HFS 100	Human Factors Engineering of Video Display Terminal Work Stations

properties of dress ophthalmic lenses. The ANSI Z80.3 standard on nonprescription sunglasses specifies the minimum UV protective properties of over-the-counter sunglasses. Table 12.1 shows the various categories described under this standard.

Most standards-setting bodies in the United States are voluntary, non-governmental groups. Because the standards they develop usually have an effect upon trade, commerce and the public good, any group that can be affected by the establishment of a standard can participate in the development of that standard. Standards organizations require that a consensus of those participating in the standard agree upon the standard before it is adopted. Since these standards-setting bodies are voluntary organizations, the standards by themselves do not establish governmental policy or law and are therefore unenforceable. However, since all interested parties participate in the development of the standard, and consensus is required for the adoption of a standard, general voluntary adherence to the standard usually occurs.

One standard which is of interest to the ergonomic community is the ANSI/HFS 100-1988. This is the American National Standard for Human Factors Engineering of Visual Display Terminal Workstations, which was published in 1988 by the Human Factors and Ergonomic Society (HFES). Although it is currently undergoing revision, it is still a valid standard for the evaluation of VDTs, workstations and lighting requirements. A copy of this can be obtained from the HFES at their address listed in the resource guide at the end of this book.

In some cases, standards that have been adopted by a voluntary standards organization become mandatory when they are adopted by a governmental agency. This occurred when the Food and Drug Administration (FDA) adopted the impact-resistant testing of ophthalmic lenses as described in ANSI Z80.1. When this became FDA policy, all dress ophthalmic lenses dispensed in this country had to meet this standard. Similarly, since OSHA adopted ANSI Z87.1 as government policy, all industrial safety eyewear must meet that standard.

Numerous other organizations establish standards, however, none of these standards organizations have impacted the field of ophthalmics as greatly as has ANSI. The American Society of Testing and Materials (ASTM) is a standards-setting organization that has numerous committees that work on standards in particular areas of industry. It has been particularly active in setting standards for the testing and quality of materials used in industry. ASTM has developed some standards for sports eye protective equipment that have subsequently been adopted by ANSI.

The Illuminating Engineering Society (IES) sets standards for lighting including those related to the quality and quantity of light in various occupational, recreational and home environments. The IES publication RP-24 is the IES Recommended Practice for Lighting Offices Containing Computer Visual Display Terminals. This standard was published in 1987 and is also a valid standard for office lighting in the VDT environment. A more complete list of vision-related standards and standards organizations is available from the American Optometric Association (Sheedy and Chioran, 1981).

The International Standards Organization (ISO) has the largest influence on ophthalmic standards. ISO standards cover all technical fields except electro-technical standards. The ISO was established in 1946 by the national standards association from 25 countries. ANSI was the founding member organization from the United States and continues as a member today. The work of ISO is carried out by 164 Technical Committees (TC). The TC of greatest interest to the ophthalmic community is ISO/TC 172, the Technical Committee on Optics and Optical Instruments. TC 172 is divided into sub-committees, each of which has several working groups to develop the specific standards documents. The participating nations in the work of ISO/TC 172 are Australia, Austria, Canada, France, Germany, Italy, Japan, United Kingdom, United States and Russia.

The standards established by ANSI, ISO and the other standards setting organizations are generally related to products or industrial application. These groups, however, are not the appropriate bodies to establish *clinical* testing standards. It is necessary that any clinical testing standards be established by clinicians because the manufacturing groups do not have clinical expertise nor should they be directly involved in the relationship between a doctor and their patient.

The current state of ergonomic legislation

Judging by the name of this section, it is likely that this information will probably be obsolete by the time you are reading it. That is because the *current* state of legislation revolving around ergonomics is in an almost constant state of flux. As soon as one piece of legislation has been enacted, it has been challenged and repealed by the courts. Even the Federal OSHA standards are dependent on the current make up of the Congress, which changes every two years. So it should come as no surprise that there is plenty of confusion as to what is required by the law regarding ergonomics.

First, though, it is important to differentiate between the various degrees of legislation which affects our business policies. As discussed above, a *standard* is simply a voluntary set of guidelines which may or may not be adopted as legal policy. Standards have the same status as guidelines – they are non-binding. However, a *regulation*, as dictated by a Regulatory Committee, has the effect of law and is a binding guideline. Likewise, a piece of *legislation* is passed by a governmental body and carries the binding effect of law.

Not every piece of legislation, however, has control over all ergonomic issues. This point was no more obvious than when Suffolk County in New York passed a law which mandated certain restrictions and accommodations for computer users. The legislation was later overturned by the courts, who indicated that it was not in the jurisdiction of the county to pass such a law. The process was nearly repeated in San Francisco, California where a similar law was passed by the city council. This resolution was also overturned by the courts, citing a similar reason as the previous legislation. It seems that only State and Federal governments have the legal right to pass laws which protect computer users.

So it fell on Federal OSHA to attempt to put together a palatable ergonomic law that would appeal to the employee, yet allow small businesses to maintain economic viability. The goal of OSHA, in general, can be summarized this way:

> *To create a safe workplace is central to our ability to enjoy health, security, and the opportunity to achieve the American dream. Accordingly, assuring worker safety in a complex and sometimes dangerous modern economy is a vital function of our government. Since OSHA was created in 1970, the agency's mission has been clear and unwavering for nearly twenty-five years: 'to assure so far as possible every working man and woman in the nation safe and healthful working conditions.' That mission – to save lives, prevent injuries and illnesses, and to protect the health of America's workers – remains vital today. (OSHA, 1996)*

At one time, OSHA might have promulgated a detailed, lengthy specification standard to address ergonomics resulting in rigid and inflexible requirements. Today, in cooperation with business and labor, OSHA will build and implement a sensible and versatile strategy for the control of work-related musculoskeletal disorders. This effort will reward exemplary employers, recognize employers requesting assistance and fairly address employers who fail to keep workplaces free of recognized and serious ergonomically-related hazards. Among other methods, this effort will spotlight those participants in OSHA's Voluntary Protection Program with exemplary ergonomic programs and emphasize their successful approaches. OSHA consultation programs in every State will promote a special program for ergonomic cooperative programs that includes outreach and program development assistance. A training and education effort including grants, seminars and materials will be coordinated with labor and industry groups. An enforcement and litigation strategy that uses the range of enforcement interventions from non-formal resolution to flexible settlement to strong enforcement will be designed to leverage OSHA's limited resources to focus on the most hazardous workplaces.

Given that, we find that OSHA has no ergonomics issues on its current agenda, as of this writing. This has obvious political implications which is beyond the scope of this book. However, it is obvious that, at least on a national level, there will be no more than a smattering of voluntary efforts and general recommendations to maintain an ergonomic standard in business.

The State of California, which has the reputation for exceptional laws, has proposed a legislative standard for ergonomics. However, in the past two years since the inception of the process, it has been diluted by various interests so as to be similar to the Federal plan: vague and non-intrusive. What started out as a 50+ page document has been pared down to one page. Here is a summary of its contents:

This section shall apply to a job, process, or operation of identical work activity at the workplace where repetitive motion injuries (RMIs) occur after [effective date]. For purposes of this section, RMIs are injuries resulting from a repetitive job, process, or operation of identical work activity at the workplace which have been the predominant cause of a diagnosed, objectively identified, musculoskeletal injury to more than one employee within the last 12 months. The diagnosis of an RMI shall be performed by a licensed physician. For definition purposes, predominant means 50% or more of the injury was caused by a repetitive job, process or operation of identical work activity. Exemption: Employers with 9 or fewer employees.

The proposed standard then recommends that every covered employer '*implement a program designed to minimize RMIs. The program shall include a worksite evaluation, control of exposures which have caused RMIs and training of employees.*' In the area of remediation, the standard suggests that '*any exposures that caused RMIs shall, in a timely manner, be corrected or if not capable of being corrected have the exposures minimized to the extent feasible. The employer shall consider engineering controls, such as work station redesign, adjustable fixtures or tool redesign, and administrative controls, such as job rotation, work pacing or work breaks.*' The complete standard is listed in Appendix G.

In its current form, the Cal-OSHA standard has moved from a proactive to a reactive position. While some companies have expressed concern about the economic impact of the proposed standard, others are fully supportive and are taking a more proactive approach. It would be nice if the general wording of the standard would be sufficient to have employees covered with regard to RMIs and still have the backing of the employer. Only time will tell if this balance has been achieved. If this standard reaches the level of law, then many other states might look toward it with an eye on their own ergonomic standards.

The EC directives

One place where the area of ergonomics has already been incorporated into the laws is in the European Community (EC). On 27 May 1990, the European Commission adopted a directive on 'the minimum safety and health requirements for work with display screen equipment'. After more than two years in incubation, the Video Display Unit (VDU) directive finally became law in Europe. Here is a summarized version of the contents of the directive with a look at its likely impact on those who design, supply, sell, buy, use or manage display screen equipment. The directive represents a major step forward for ergonomics. The preamble, amongst other things, obliges employers to keep themselves informed of 'scientific findings concerning workstation design' and stresses that the ergonomic aspects are of particular importance for a workstation with display screen equipment. The document is structured as a series of 12 articles organized in three sections.

SECTION ONE: GENERAL PROVISIONS
Article 1: Subject
The original text had few exclusions and covered 'all workstations equipped with a VDU'. Successful lobbying from various sources resulted in control cabs, portable systems, public terminals, calculators, cash registers and typewriters being excluded. However, the scope still remains broad and is likely to be just one of the issues which will continue to be debated long after the directive is implemented.

Article 2: Definitions
This article makes it clear that the directive applies to the entire workstation including 'display screen, keyboard, peripherals – including diskette drive, printer, document holder, work chair, work desk and the immediate work environment'. The definition of worker, for the purposes of the directive, includes anyone who 'habitually uses display screen equipment as a significant part of his normal work'.

SECTION TWO: EMPLOYER'S OBLIGATIONS
Article 3: Analysis of workstations
This article obliges employers to evaluate VDU health and safety and to take appropriate measures to remedy the risks found, especially with regard to eyesight, physical problems and mental stress. In line with other regulations, this places the initiative firmly on the employers of VDU workers. Ignorance is no excuse.

Article 4: Workstations put into service for the first time
Workstations first put into service after 31 December 1992 must meet the minimum requirements laid down in the technical annex. This represents a significant tightening of the directive since its first version allowed two years after the directive came into force before it became mandatory.

Article 5: Workstations already put into service
Allows four years for existing workplaces to be adapted to comply. This too has been tightened compared to the previous formulation which included the phrase 'so far as is reasonably possible'. The second version makes no such concession.

Article 6: Information for, and training of workers
In addition to general guidance on VDU health and safety, workers and/or their representatives shall also be informed of the outcome of the analysis of the workplace carried out under Article 3. It will no longer be sufficient for employers to assess risk and decide on appropriate action. This article will make it essential to involve staff in the process and will require employers to inform them of any risks involved in the workplace. In the United Kingdom, the Health and

Safety at Work Act already makes safety everyone's responsibility. This article reinforces the manager's role in anticipating risks and dealing with them appropriately in co-operation with staff.

Article 7: Daily work routine
This places a responsibility on the employer to plan workers' activities in such a way that daily work on a display screen is periodically interrupted by breaks or by changes of activity.

Article 8: Worker consultation and participation
This article provides a mandate for worker participation in determining VDU health and safety issues within organizations. This is in line with the Social Charter and out of line with current UK Government's approach to industrial relations.

Article 9: Protection of workers' eyes and eyesight
This directs that workers shall be entitled to a sight test before VDU use regularly and if they experience visual difficulties. If this test warrants it they shall also be entitled to an ophthalmologic or optometric examination. It also requires that if special glasses are necessary, the employer shall provide them. None of this protection is to involve the employee in financial cost.

VDU eye tests were an important negotiating point in the United Kingdom some years ago, but declined in importance once everyone realized that regular free eye tests were part of the National Health Service and generally 'a good thing'. The reintroduction of eye test charges has raised the profile of this issue and the directive puts it firmly back on the agenda.

SECTION THREE: MISCELLANEOUS PROVISIONS
Article 10: Adaptations to the annex
This allows the annex to be updated taking account of technical developments in international regulations and specifications.

The implications of the directive has caused considerable concern in the computer industry. It has also concerned organizations which have large numbers of staff using VDUs. Given the potential impact of the directive in Member States, this concern is clearly justified. There is now legal support in Europe for ergonomics addressing all aspects of the user interface from VDU hardware to software and dialogue issues. This will bring the issues into much sharper focus than at present. Employers who ignore these issues will open themselves to expensive litigation.

However, it is not likely that good employers have as much to fear as the publicity suggests. Despite the release, many of the requirements of the directive make good ergonomic sense and are in line with what has been recommended for several years.

Reference

Sheedy, J.E. and Chioran, G.M. (1981) *Vision Standards Bibliography*. St. Louis: American Optometric Association.

Chapter 13

Epilogue

Technology continues to advance with the supposed purpose of enhancing the way we live our lives. Yet, we – as human beings – are left to adapt to this new way of dealing with our world. We marvel at the abilities this new technology has afforded us – from traveling to new worlds to exploring new uses for the atom – and try to grasp the meaning of it all. Yet, we are still humans in our 'original' physical form and we must first consider how we can benefit from our new 'tools'.

Our primary way of interacting with our environment is through our sense of vision. It accounts for over 80% of our learning and it is the sense we fear losing the most. How we see literally affects the quality of our lives. And this sense of vision doesn't reside in our eyes – they are just the receivers of the light. The true sense of vision lies within, and throughout, our brain – the control center of our body.

The Information Age, with computers as the core technology, has dawned. The computer has been called 'the machine that changed the world' and it has certainly earned that label. It has become our preferred method of communication and has effectively made the world a smaller place in which to interact. There seems little doubt that this trend will continue but I will leave it to the visionaries to predict how it will evolve. Yet, our dependence on the computer will continue to deepen. What form that will take is still to be determined, but it is apparent that our vision will continue to play a significant role in this interaction. No matter what type of device is used to input information into the computer, it will always be our eyes which deal with the input.

It is apparent that, for the most part, we realize what is happening, yet changing our habits is a slow and arduous process. The awareness of VDT worker discomfort and the willingness to invest in office workplace improvements to help control the continual problems is still a low priority for many companies. It is clear that employees can 'flounder or flourish' depending upon the accommodations for working comfort afforded them. Employers should be concerned because they pay the price in health care costs, workers' compensation claims, lowered productivity, absenteeism and other indemnities that poor working conditions create. Employers must take action today, or face the consequences tomorrow for any indifference in the face of vision and physical disorders increasingly experienced by office workers.

How will our eyes adapt to this new and stressful mode of use? So far, it appears that our visual system will adapt in one of three possible ways. For some of us, our eyes adapt by changing to see clearer at this near point of vision, becoming near-sighted. This adaptation makes it easier to see clearly up close, despite losing our clarity for distance viewing. This is currently one of the more common adaptations. Secondly, there are those who will experience physical problems due to non-adaptation. These symptoms include headaches, sore or red eyes, even neck and shoulder pains. It will be difficult for these people to deal with this new technology and may follow other paths of work to avoid the computer experience. The third type of adaptation is the succesful one. It will entail the use of occasional near viewing and alternate distance viewing which can balance out the visual system. This alternative viewing habit should also include physical breaks so that the body, as well as the eyes, receive an alternative task to perform.

Unfortunately, there is no 'magic bullet' which will address all of the various issues in every workplace. This book is intended to be a starting point to begin an inquiry into resolving many of these issues. It is important to remember that the vision problems which affect VDT users are a combination of environmental conditions, work habits and visual system characteristics. Therefore, it is critical that the visual abilities of the VDT worker be tested on a continual basis to assure that they can perform their duties adequately and productively.

Our system of eye examinations must also change to keep up with our new way of interacting with our environment. Currently, the eye care professions use techniques of examination which are at least 60 years old. Yes, there are machines which make use of current technology which can increase the efficiency of the eye examination procedure. However, it will eventually be a return to the past – an evaluation of how we function in our environment – which must be addressed in the eye exams of the future. Our eyes are being used at a new working distance and the visual posture at that distance must be addressed. The subspecialty of Environmental Optometry is one which will be expanding simply based on the need to address the concerns presented in this book.

It is hoped that the information presented here will spark the interest of the employee and management so they can work together to maintain a healthy, happy and productive workplace. The answer to many problems may be right before your eyes!

Appendix A

VDT Workplace Questionnaire

Work Practices:
1. Number of hours per workday of VDT viewing. _____
2. How long have you worked at a VDT job? _____
3. Type of work habits: a) Intermittent – periods of less than 1 hour
   (circle one)   b) Intermittent – periods of more than 1 hour
                           c) Constant – informal breaks, as required
                           d) Constant – regular breaks
                           e) Constant – no breaks, other than meals
4. How often do you clean your display screen? _____

Environment:
Lighting in the work area: (check all that apply)
Fluorescent overhead only _____
Incandescent overhead only _____
Fluorescent and incandescent overhead _____
Fluorescent overhead and incandescent direct _____
Window light _____ In front? Behind? To the side?
Window light control: Curtains? Blinds? Vertical/Horizontal?
Desk Lamp/Task Light _____
Other (describe) _____
Walls: Color _____ Shiny / Dull finish?
Desk surfaces: Color _____ Shiny / Dull finish?
How would you rate the brightness of the room: Very bright / Medium / Dim?

Display Screen:
What color are the letters on your screen? _____
What color is the background of your screen? _____
Viewing distance from your eye to VDT screen: _____ inches.
Can the monitor be tilted? Y N
Can the monitor be raised / lowered?
Do you notice the screen flicker? Y N
Does the screen have a glare filter? Y N If so, is it glass/mesh?
Top of VDT screen (above, equal to, below) eye level?
If above or below, by how many inches? _____

Workstation:
Viewing distance from your eye to keyboard: _____ inches.
Viewing distance from your eye to hard copy material: _____ inches.
Reference material is (to the side, below) the screen?
If to the side, is it next to the screen or keyboard? Is this height adjustable? Y N
Is the monitor supported on a (stand / desk / CPU)?
Is this adjustable? Y N
Is all of your hard copy material visible without significant movements? Y N

Visual Symptoms:
Do you experience any of the following symptoms during or after VDT work?:
- ❑ Eyestrain
- ❑ Headaches
- ❑ Blurred Near Vision
- ❑ Blurred Distant Vision
- ❑ Dry / Irritated Eyes
- ❑ Double Vision
- ❑ Neck / Shoulder / Wrist ache
- ❑ Color Distortion
- ❑ Light Sensitivity
- ❑ Backache

Do you wear glasses while working at the VDT? Y N
If yes, are they (single vision, bifocal or progressive)?

Do you wear contact lenses while working at the VDT? Y N
If yes, are they (soft, gas permeable, or hard lenses)?

Appendix B

Occupational Vision Requirements Questionnaire

Employee _____
Employer _____

The doctor needs the following information for the analysis of your occupational vision needs, and for the design of appropriate occupational/protective eyewear.
Brief job description_____
_____

Please check each item related to your job:
Work is performed while:
☐ Standing ☐ Sitting ☐ Walking ☐ Driving
☐ Indoors ☐ Outdoors ☐ Viewing a video screen

Seeing directions and distances:
☐ Eye level   Viewing distance is _____ inches/feet.
☐ Below eye level   Viewing distance is _____ inches/feet.
☐ Above eye level   Viewing distance is _____ inches/feet.
☐ Other directions and distances (*please describe*)
_____

Viewing area (field of view) at:
Far distance is   ☐ Large ☐ Medium ☐ Small
Intermediate distance is ☐ Large ☐ Medium ☐ Small
Near distance is   ☐ Large ☐ Medium ☐ Small

Size of tasks at:
Far distances   ☐ Large ☐ Medium ☐ Small ☐ Very Small
Intermediate distances ☐ Large ☐ Medium ☐ Small ☐ Very Small
Near distances   ☐ Large ☐ Medium ☐ Small ☐ Very Small

Work environment:
Temperature: ☐ Hot   ☐ Cold ☐ Average
Lighting: ☐ Bright ☐ Dark ☐ Average
Humidity: ☐ High   ☐ Low   ☐ Air conditioned
Other (*please describe*) _____

Eye hazards:
☐ Metal particles ☐ Non-metal particles ☐ Dust
☐ Fumes   ☐ Chemical splash   ☐ Moving machinery
☐ Infra red   ☐ Ultra violet   ☐ Glare   ☐ Radiation
Other (*please describe*) _____

Special vision requirements:
☐ Depth perception _____
☐ Color discrimination _____
☐ Other (*please describe*) _____

Please present this completed report to the doctor you select.

Appendix C

Resources for the blind and visually impaired

This list contains useful sources of information for blind and visually impaired computer users. The list is divided into four sections: Organizations, Newsletters and Journals, Networks, Databases and Bulletin Boards, and Books and Pamphlets. Please note that the list is based on information used in research and outreach programs. It is therefore likely that the list is not exhaustive. No recommendations or endorsements are implied by inclusion on this list.

Organizations

AFB Technology Center
American Foundation for the Blind, Inc.
11 Penn Plaza
Suite 300
New York, NY 10001
(212) 502-7642
(212) 502-7773 (FAX)
email: techetr@aft.org

Conducts product evaluations of assistive technology including Braille technology, optical character readers, speech Trimesters, screen magnifiers, and closed circuit televisions

American Council of the Blind
1155 15th Street, Suite 720
Washington, DC 20005
(202) 467-5081
(202) 467-5085 (FAX)

Publishes computer resource list about various devices and where to buy them. Visually Impaired Data Processors International, a computer users' special interest group, is part of ACB.

American Printing House for the Blind
PO Box 6085
1839 Frankfort Avenue
Louisville, KY 40206
(502) 895-2405
(800) 223-1939

Producers of software for users with visual impairments. APH also produces user manuals in Braille for Apple computers, instructional aids, tools, and supplies. Four-track recorder / players available. Tutorial kit for Microsoft Windows available for VI users.

Braille Institute
741 N. Vermont Avenue
Los Angeles, CA 90029
(213) 663-1111

This is an educational organization for persons who are blind, deaf-blind, or partially sighted. It has technological resources, a talking book library, and a large community outreach program. A subsidy program for funding equipment is available to persons who are currently employed and legally blind.

Carroll Center for the Blind
70 Centre Street
Newton, MA 02158
(617) 969-6200

Various publications and rehabilitation and educational programs are available for persons who are blind or visually impaired. A computer training program, Project CABLE, provides computer assessment, training on adaptive devices and software, and word processing training. Summer training courses for youth are also offered.

Central Blind Rehabilitation Center
Veterans Affairs
Edward Hines, Jr. VA Hospital
PO Box 5000 (124)
Hines, IL 60141-5000
(708) 216-2271

Information on various types of computer access devices for people with visual impairments.

Centre for Sight Enhancement
School of Optometry
University of Waterloo
Waterloo, ON N2L 3GI
CANADA
(519) 8894708
(519) 746-2337 (FAX)

A clinical teaching, and research facility providing assessment, prescription, instruction and/or rehabilitation by a multidisciplinary professional team. Provides sight enhancement devices under the Provincial Ministry of Health's Assistive Devices Program.

Helen Keller National Center for Deaf-Blind Youths and Adults
111 Middle Neck Road
Sands Point NY 11050 (516) 944-8900
(516) 944-8637 (TTY)

Only national program that provides diagnostic evaluation. short-term comprehensive rehabilitation and personal adjustment training, and job preparation and placement for Americans who are deaf-blind. Local services provided through regional offices, affiliated agencies, a National Training Team, and a Technical Assistance Center for older adults.

International Braille and Technology Center for the Blind
National Center for the Blind
1800 Johnson Street
Baltimore, MD 21230
(410) 659-9314

Provides demonstrations, comparative evaluations, cost comparison, ADA compliance assistance, personal and telephone consultation pertaining to assistive technology for the visually impaired, tours meeting and conference facilities. Resource for blind persons and

Appendix C

sighted individuals. Overnight and dining accommodations may be available for a fee.

National Association for Visually Handicapped
22 West 21st Street
New York NY 10010
(212) 889-3141

Organization dealing with the needs of people who are partially sighted. Contact for information about computer access. Also a San Francisco office at 3201 Balboa Street, San Francisco CA 94121.

National Braille Press, Inc.
88 St. Stephen Street
Boston, MA 02115
(617) 266-6160

Publications on personal computer technology for people who are blind. Many printer and modem manuals transcribed in Braille.

National Federation of the Blind
1800 Johnson Street
Baltimore, MD 21230
(410) 659-9314

Programs include: Committee on Evaluation of Technology (which evaluates current and proposed technology for people who are blind or visually impaired); International Braille and Technology Center for the Blind (a demonstration and evaluation center for computer technology for blind and visually impaired users): and NFB in Computer Science (a nationwide computer users' group which publishes an annual newsletter for people who are blind or visually impaired).

Newspapers For the Blind
DataCast Communications, Inc.
900 Lady ellen Place Suite 23
Ottawa, ON KIZ 5L5
CANADA
(613) 725-2106
(613) 722-8756 (FAX)

Service that delivers an electronic version of Canada's well known newspapers to computers via a VBI decoder attached to cable TV. Available newspapers are: Montreal Gazette, La Presse, Ottawa Citizen, Toronto Star, Toronto Sun, Financial Post, The Globe and Mail.

Recording for the Blind and Dyslexic (RFBD)
20 Rozel Road
Princeton, NJ 08540
(800) 221-4792
(609) 452-0606

Provides academic textbooks and other educational textbooks to people who cannot read standard print because of physical, perceptual or other disabilities. Must be a registered member in order to borrow material; call for details. Also sells reference books on disk and related software products.

Sensory Access Foundation
385 Sherman Avenue, Suite 2

Palo Alto, CA 94306
(414) 329-0430 (voice)
(415) 329-0433 (TDD)

Compiles and publishes consumer information on technology updates, including computer adaptations, for blind and visually impaired people. Assists in career placement for people who have visual impairments. Career services are only available within California: information services available worldwide. Publishes a magazine and has a technology training center.

Smith-Kettlewell Eye Research Institute
Rehabilitation Engineering Center
2232 Webster Street
San Francisco, CA 94115
(415) 561-1619

Research and development center on assistive technology (including work on computer access devices) for people who are blind or visually impaired.

Technology Center (TC)
American Foundation for the Blind
11 Penn Plaza
Suite 300
New York NY 10001
(212) 502-7642
(800) 232-5463
E-mail to: techtr@afb.org.
Gopher: gopher@afb.org

Conducts evaluations of assistive technology for visually impaired people and provides information on those products. Coordinates the Careers and Technology Information Bank (CTIB), a collection of data from visually impaired people who use adaptive equipment in a variety of jobs.

Newsletters and journals

Braille Forum
American Council of the Blind
1155 15th Street
Suite 720
Washington, DC 20005
(202) 467-5081
(202) 467-5085 (FAX)

Publication dealing with blindness-related issues, such as legislation, technology, and product and service announcements.

Computer Folks
c/o Richard Ring
269 Terhune Avenue
Passaic, NJ 07055
(201) 471-4211

A magazine on cassette for blind computer users by blind computer users. Discusses and demonstrates adaptive technology.

Appendix C

Journal of Visual Impairment and Blindness
American Federation for the Blind
11 Penn Plaza
Suite 300
New York, NY 10001
(212) 502-7600

Research journal on issues related to visual impairment and blindness. Includes research reviews, application papers, and articles on special topics (including assistive technology). Published monthly, except July and August.

Tactic
Clovernook Center
7000 Hamilton Avenue
Cincinnati, OH 45231
(513) 522-3860

International quarterly offers information and reviews on technology for people with visual impairments. Published in Braille, large print, and diskette (IBM compatible) formats.

Technology Update
Sensory Access Foundation
385 Sherman Avenue Suite #2
Palo Alto, CA 94306
(415) 329-0430 (voice)
(415) 329-0433 (TDD)

Bimonthly newsletter with information regarding technology and vision impairment. Includes new product announcements, product reviews, and consumer information. Available in print, large print, cassette, and PX diskette.

Visual Field
Florida Instructional Materials Center
5002 North Lois Avenue
Tampa, FL 33614
(813) 872-5281
(813) 872-5284 (FAX)

Biannual newsletter on products, projects, conferences, etc. related to education of students with visual impairments.

Networks, databases and bulletin boards

4 Sights Network
Upshaw Institute for the Blind
16625 Grand River
Detroit, MI 48227
(313) 272-3900
(314) 272-7111 (dial in)

This computer network provides bulletin board and database information for blind and visually impaired individuals and those working with them. The information covers vocational and rehabilitation resources, assistive technology, educational information for parents, teachers, and students, public policy and more.

Carl et Al
American Printing House for the Blind
Attn: Paul Brown
PO Box 6085
Louisville, KY 40206-0085
(800) 223-1939
(502) 895-1509 (FAX)

An on-line database that lists materials in media accessible to people who are visually impaired. Over 120,000 records including Braille books, large type materials, music scores, electronic books, sound recordings, software programs, and tactile graphics. Contact APH for billing and access information.

Books and pamphlets

CD-ROM Advantage
D. Croft, D. Kendrick, and A. Gayzagian 1994
National Braille Press
88 St. Stephen Street
Boston, MA 02115
(617) 266-6160
(617) 437-0456 (FAX)

Answers commonly asked questions about CD-ROM technology and how it works with speech and Braille. Includes practical advice from users, profiles blind users of CD-ROM, and lists over 100 CD-ROM titles.

Computer Access, Resource Manual
Rosenbaum, et al 1987
Carroll Center for the Blind
770 Centre Street
Newton, MA 02158

Resource manual and curriculum for setting up an evaluation and training center in assistive technology applications for blind and visually impaired.

Customer Service Representative Training Manual
Ferrarin, Rosenbaum, et al.
1994
Carroll Center for the Blind
770 Centre Street
Newton, MA 02158

A comprehensive curriculum for creating and providing job readiness skills for employment in customer service jobs to individuals with visual impairments utilizing assistive technology. For rehabilitation agencies, secondary institutions, or career counselors.

Extend Their Reach
1990
Electronic Industries Association
2500 Wilson Boulevard
Arlington, VA 22201
(703) 907-7600

Gives introduction to the types of products available to overcome impairments of sight, speech, hearing, motion, etc. Also provides

Appendix C

information on how to find funding and producers of assistive devices, listing of companies, and resources for further information.

Job Readiness Workshop
Carroll Center for the Blind
770 Centre Street
Newton, MA 02158
(617) 969-6200

Step by step guide on creating a résumé and developing interviewing techniques. Sample résumés and example interview questions given.

Medical Transcription Training Manual
Ferrarini, Fosenbaum, et al.
1994
Carroll Center for the Blind
770 Centre Street
Newton, NM 02158
(617) 969-6200

Curriculum, guidelines, and resources for creating or modifying medical transcription programs for persons with visual disabilities. 180-page manual includes detailed course outlines for terminology, transcription and computer instruction, resources, job information.

Print and Braille Literacy: Selecting Appropriate Learning Media
Hilda Caton, Ed.D.
American Printing House for the Blind, Inc.
1839 Frankfort Avenue
PO Box 6085
Louisville, KY 40206
(502) 895-2405

Provides guidelines designed to ensure that every visually impaired student will have adequate opportunity for learning to use the medium/media most appropriate for his or her needs. Guidelines were developed by a committee of experts in the field.

Project CABLE Resource Manual, 2nd edition 1987
Project CABLE (Computer Access for the Blind in Education and Employment)
Carroll Center for the Blind
770 Centre Street
Newton, MA 02158
(617) 969-6200

This manual includes curriculum, evaluation form, lesson plans, and other resources to assist in setting up or running an employment program for persons who are blind or visually impaired. Other topics include funding and staffing.

TeleSelling Training Manual
Carroll Center for the Blind
770 Centre Street
Newton, MA 02158
(617) 969-6200

Curriculum, guidelines, and resources for creating, or modifying a telemarketing training program for persons with visual disabilities. Chapters include information about the field, developing a program, teleselling skills, computer course outlines, job readiness workshop.

Tools for Selecting Appropriate Learning Media
American Printing House for the Blind
PO Box 6095
1839 Frankfort Avenue
Louisville, KY 40206-0085
(800) 223-1839

An extension of the book Print and Braille Literacy. *Designed to help parents, teachers and administrators make decisions concerning students' use of Braille, print, or both as their primary reading medium/media.*

Vendor Information Sheets
Consumer Information Department of Sensory Access Foundation
385 Sherman Avenue, Suite 2
Palo Alto, CA 94306
(415) 329-0430
(415) 323-1062 (FAX)

Lists vendor names, addresses, phone numbers, fax numbers, BBSs and prices for access products for people who are blind or have low vision. Sheets available for: screen readers, speech synthesizers, computer magnification, closed circuit televisions, Braille devices and software, reading machines, and others.

Appendix D

Computer access products for blind and visually impaired users

Listed below are makers of products which allow blind and visually impaired people to use computers. The three types of output which these systems use are: synthesized speech, large print, and Braille or other tactile output. The products listed may consist of hardware, software, or a combination. Braille printers are also included in this list. Most of these products are designed to provide access to standard commercial software (word processors, spreadsheets, etc.) designed for the non-disabled marketplace. A few are dedicated programs, specifically designed for people with visual impairments but not providing access to other software.

These addresses are provided so you can find out what is available. We suggest you contact the companies for more information, and that you also do some investigating on your own, such as reading books or magazines, talking to users of the products or consulting experts on the topic.

Access Systems International, LTD
415 English Avenue
Monterey, CA 93940
(408) 375-5313
   Braille Access
   Index Basic
   Index Basic-D
   Index Everest-D

ADA Compliance Information
US Department of Justice
800-514-0301
800-514-0383 (TDD)
http://www.usdoj.gov/crt/ada/adahom1.htm

Al Squared
P.O. Box 669
Manchester Center, VT 05255-0669
(802) 362-3612
FAX (802) 362-1670
   ZoomText 5

American Printing House for the Blind (APH)
P.O. Box 6085
Louisville, KY 40206-0085
(502) 895-2405
(800) 223-1839
   Braille 'n Speak Classic
   Echo II with Textalker & Echo II with Textalker-GSNOMAD Talking Touch Pad & NOMAD Gold
   Speaqualizer Speech Access System
   TEXTALKER
   Textalker-GS

American Thermoform Corporation
2311 Travers Avenue
City of Commerce, CA 90040
(213) 723-9021
   Braille 200
   Braille 400 S
   Braille Comet
   KTS Braille Display
   Ohtsuki BT-5000 Braille/Print Printer

Apple Computer, Inc.
Worldwide Disability Solutions Group, MS 38DS. I Infinite Loop
Cupertino, CA 95014
(800) 600-7808
TT/TDD (800) 755-0601
applesdsg@eworld.com
Http://www.apple.com/disability/
   CloseView

Arkenstone, Inc.
1390 Borregas Avenue
Sunnyvale, CA 94089
(800) 444 4443
(408) 752-2200
TT/TDD (800) 833-2753
info@arkenstone.org
   Open Book
   Open Book Unbound

Artic Technologies
55 Park Street
Troy, MI 48083
(810) 588-7370

FAX (810) 588-2650
  Artic TransBook
  Artic TransType
  Business Vision
  Gizmo
  Magnum Deluxe
  Magnum GT
  Win Vision

ATR/JWA, Inc.
P.O. Box 180
Fairfax Station, VA 22039
(703) 715-6072
FAX (703) 903-9142
  BRAILLEX-2D
  Notex 486

Berkeley Systems, Inc.
2095 Rose St.
Berkeley, CA. 94709
(510) 540-5535 ext. 716
TT/TDD (510) 849-9426
osw@berksys.com
  inLARGE 2.0
  outSPOKEN 1.7
  outSPOKEN for Windows

Biolink Computer Research
and Development, LTD
Suite 105 - 140 West 15th
North Vancouver, BC
V7M IR6 CANADA
(604) 984-4099
BBS (604) 985-8431
FAX (604) 985-8493
  Protalk23 for Windows

B-K Press of Texas
P O Box 4843
Wichita Falls, TX 76308
817-723-0254 FAX
Http://www.theshoppes.com/~bkpress

Blazie Engineering
105 East Jarrettsville Road
Forest Hill, MD 21050
(410) 893-9333
BBS (410) 893-8944
FAX (410) 836-5040
info@blazie.com
WWW: http://blazie.com/
  Braille 'n Speak
  Braille Blazer
  Braille Lite
  Type 'n Speak

Carolyn's
Box 14577
Bradenton, FL. 34820-4577
800-648-2266/FAX 813-761-8306

Compusult Limited
40 Bannister Street
P.O. Box 1000
Mount Pearl, NF AIN 3C9 CANADA
(709) 745-7914
FAX (709) 745-7927
scantell@compusult.nf.ca
http://www.compusult.nf.ca
  ScanTELL

Duxbury Systems, Inc.
435 King Street
Littleton, MA 01460
508-486-9766

Easier Ways, Inc.
1101 N. Calvert Street, Suite 405
Baltimore, MD. 21202
410-659-0232 / 410-659-0233 (FAX)

Enabling Technologies Company
3102 S.E. Jay Street
Stuart, FL 34997
(800) 777-3687
FAX (407) 220-2920
  Braille BookMaker and Braille Express 100 &
  500
  Juliet Brailler
  Marathon Brailler
  Romeo Brailler Model RB-25
  Romeo Brailler Model RB-40
  Thomas Brailler
  TranSend System

Equipment for Visually Disabled People: An
International Guide
Technical Research and Development
Department
Royal National Institute for the Blind
224 Great Portland Street
London WIN 6AA, England

Florida New Concepts Marketing, Inc.
P.O. Box 261, Port Richey, FL 34673
(813) 842-3231
FAX (813) 845-7544
  Beamscope
  Compu-Lenz

Franklin Electronic Publishers, Inc.
Franklin Learning Resources Division

Appendix D

One Franklin Plaza
Burlington, NJ 08016-4907
(800) 525-9673
(609) 386-2500
   Language Master Special Edition
   LM6000SE

General Services Administration
Center for Information Technology
Accommodation
18th & F Street, NW
Room 1234
Washington, DC 20405
202-501-4906 (Voice)
202-501-2010 (TTY)/ 202-501-6269 (FAX)

GW Micro
310 Racquet Drive
Fort Wayne, IN 46825
(219) 483-3625
FAX (219) 482-2492
support@gwmicro.com
   Vocal-Eyes
   Window-Eyes

Henter-Joyce, Inc.
2100 62nd Avenue North
St. Petersburg, FL 33702
(800) 336-5658
(813) 528-8900
BBS (813) 528-8903
FAX (813) 528-8901
7477.3306@compuserve.com
   JAWS for Windows
   JAWS Screenreader
   WordScholar

Hexagon Products
P.O. Box 1295
Park Ridge, IL 60068
(708) 692-3355
76064.1776@compuserve.com
   B-Pop
   Big-W

HumanWare, Inc.
6245 King Road
Loomis, CA 95650
(800) 722-3393
(708) 620-0722
FAX (916) 652-7296
   ALVA Braille Terminals
   Braille-N-Print
   Keynote Companion
   MasterTOUCH
   Mountbatten Brailler

   Paragon Braille Printer
   Ransley Braille Interface (RBI)
   Soundproof

IBM Corporation
Special Needs Systems
P.O. Box 1328
Internal Zip 5432
Boca Raton, FL 33432
(800) 426-4832
TT/TDD (800) 426-4833
   EBM Screen Reader/2
   IBM Screen Reader/DOS
   Screen Magnifier/2

Innoventions, Inc.
5921 S. Middlefield Road, Suite 102
Littleton, CO 80123
800-854-6554
Http://www.magnicam.com/magnicam/

Kansys, Inc.
4301 Wimbledon Ter. 2B
Lawrence, KS 66047
(913) 843-0351
   PROVOX

Konz, Ned
810 21st Avenue North
St. Petersburg, FL 33704
nedkonz@gate.net,
76046.223@compuserve.com
   Lens 2.03

Less Gauss Inc.
Suite 160
187 East Market Street
Rhinebeck, NY 12572
(800) 872-1051
(914) 876-5432
FAX (914) 876-2005
   Adjustable EZ Magnifier
   GNK Magnifier
   NuVu Magnifier

Lighthouse Enterprises
Consumer Products Division
36-20 Northern Blvd.
Long Island City, NY 11101
800-829-0500

LS & S Group
P O Box 673
Northbrook, IL 60065
800-468-4789 or 708-498-9777

Massachusetts Association for the Blind
200 Ivy Street

Brookline, MA 02146-3995
617-738-5110/ In Mass. 800-682-9200
TDD 617-731-6444/FAX 617-738-1247

Mayer-Johnson Company
P.O. Box 1579
Solana Beach, CA 92075
(619) 550-0084
FAX (619) 550-0084
mayerj@aol.com
   Infovox 210

Microsystems Software Inc.
600 Worcester Road
Framingham, MA 01701
(800) 828-2600
(508) 879-9000
BBS (508) 875-8009
FAX (508) 626-8515
   HandiCHAT and HandiCHAT Deluxe
   MAGic and MAGic Deluxe

MicroTalk Software
917 Clear Creek Drive
Texarkana, TX 75503
(903) 832-3471
Modem (903) 832-3722
FAX (903) 832-3722
   ASAP (Automatic Speech Access Program)

MONS International, Inc.
Products for the Visually Impaired
6595 Roswell Road #224
Atlanta, GA 30328
800-541-7903 or 404-551-8455

National Institute for Rehabilitation Engineering
P.O. Box T
Hewitt, NJ 07421
(800) 736-2216
(201) 853-6585
dons@warwick.net
   Large-Type Display Utility Software

New Concepts Marketing
P O Box 261
Port Richey, FL 34673
800-456-7097

Okay Vision-Aide Corporation
14811 Myford Rd.
Tustin, CA 92680
800-325-4488
e-mail: vision-aide@ovac.com

OMS Development
610-B Forest Avenue
Wilmette, IL 60091
(708) 251-5787
(800) 831-0272
FAX (708) 251-5793
ebholman@metcom.com
   Tinytalk

Optelec USA, Inc.
6 Lyberty Way
P.O. Box 729
Westford, MA 01886
(800) 828-1056
(508) 392-0767
   LP-DOS & LP-Windows

The Productivity Works, Inc.
7 Belmont Circle
Trenton, NJ 08618
609-984-8044
609-984-8048
http://www.prodworks.com

S. Walter, Inc.
30423 Canwood St., Suite 115
Agoura Hills, CA 91301
818-406-2202
800-992-5837

Science Products for the Blind
P O Box 888
Southeastern, PA 19399
800-888-7400

Sigma Designs, Inc.
47906 Bayside Parkway
Fremont, CA 94538
510-770-0100

Speech Systems for the Blind
76 Wheaton Drive, Attleboro, MA 02703
(508) 226-0447
73030.3644@compuserve.com
   Seekline
   WINKLINE

Syntha-Voice Computers Inc.
9009-1925 Pine Street
Niagra Falls, NY 14301
(905) 662-0565
(800) 263-4540
BBS (905) 662-0569
FAX (905) 662-0568
help@synthavoice.on.ca
ftp: synthavoice.on.ca
   Panorama
   Powerama
   Slimware
   Slimware Window Bridge

Appendix D

Technology for Independence
529 Main Street
Schraft Center Annex
Boston, MA 02129
617-242-7007

Telephone Pioneers of America
P O Box 18388
Denver, CO 80204
303-571-1200

Trace R&D Center
S-151 Waisman Center
1500 Highland Avenue
Madison, WI 53705
608-263-2309
TDD: 608-263-5408

T.V. Raman
(617) 692-7637
raman@crl.dec.com
http://www.cr.dec.com/crl/people/
  biographies/raman.html
  Emacspeak

TeleSensory Corporation
455 North Bernardo Avenue
P.O. Box 7455
Mountain View, CA 94039
(415) 960-0920
(800) 804-8004
  BrailleMate 2
  David
  DM80/FM

INKA
  Optacon II
  PowerBraille 40, 65 & 80
  ScreenPower
  ScreenPower for Windows
  Vantage CCD
  VersaPoint-40 Braille Embosser
  Vista

TFi Engineering
529 Main Street
Boston, MA 02129
(800) 843-6962
(617) 242-7007
FAX (617) 242-2007
  Myna

VisuAide
841 Jean-Paul Vincent Boulevard
Longueuil, PQ J4G IR3
CANADA
(514) 463-1717
FAX (514) 463-0120
  Magnum

Xerox Imaging Systems, Inc.
9 Centennial Drive
Peabody, MA 01960
(800) 248-6550
  BookWise
  Reading AdvantEdge

Appendix E

Seal of acceptance program for VDT glare reduction filters, July 1996

The American Optometric Association (AOA) has established a program to provide evaluation and recognition of quality video display terminal (VDT) glare reduction filters. Products that meet the minimum specifications established by the AOA Commission on Ophthalmic Standards are allowed to use the AOA Seal of Acceptance in product labeling and marketing. Specifications for VDT glare reduction filters cover construction quality, image quality, glare reduction, reflectance and the ability to withstand environmental testing,

The following is a listing of companies whose products have met the AOA specifications for VDT glare reduction. You can contact these manufacturers directly for additional information about their filters.

Optical Coating Laboratory, Inc. (OCLI)
2789 Northpoint Parkway
Santa Rosa, CA 95407
800/545-OCLI

Accepted filter models: Glare/Guard Standard, Maximum, Maximum Plus, and E-Shield; MacGlare/Guard, MacGlare/Guard Maximum Plus, and MacGlare/Guard E-Shield; Profile and Universal Profile; Optima 1500 and 1700 High Contrast and Litetint; Dr. Dean Edell anti-glare/anti-radiation screen. Accepted European filter models: MultiGuard, Cobra, MultiCADguard, Standard A/S, Standard 500 High Contrast and Litetint,

E for M Corporation, Optical Filters Group
625 Alaska Avenue
Torrance, CA 90503
310/320-9768

Accepted filter models: Vu-Tek Exec and Gold (VPND37P), Pro and Silver (VPND31 and VPND62), and Temp and Bronze (VND31).

SoftView Computer Products/ErgoView Technologies
535 Fifth Avenue, 32nd Floor
New York, NY 10017
212/867-7713 or 212/867-0661

Accepted filter models: SoftView Eye Saver, Eye Saver Plus, Contour and Contour Plus; ErgoView Visionmate, Visionmate Plus, Contour and Contour Plus, ErgoView Ultra/Fit and Ultra/Fit Plus, ErgoView Universal and Universal Plus.

3M Safety and Security Systems Division
3M Center, Bldg. 225-4N-14
St. Paul, MN 55144
800/553-9215

Accepted filter models: Anti-glare (AF 100, AF 150 and AF 200) and Anti-glare/Radiation (AF 200 and AF 250).

Fellowes Manufacturing Company
1789 Norwood Ave.
Itasca, IL 60143
708/893-1600

Accepted filter models: LiteView Glass and Contour Anti-glare, LiteView and Contour Anti-static/Radiation. Workstation Glass Anti-glare (single and double-sided coating), Workstation Anti-static/Radiation, and Notebook Anti-glare.

Kantek, Inc.
15 Main Street
East Rockaway, NY 11518
516/593-3212

Accepted filter models: Spectrum Universal and Spectrum Universal Contour.

Focus Computer Products A/S
15 Generatorvej
DK-2730 Herlev
Copenhagen, Denmark
45 4284 5144

U.S. Distributor:
Focus Computer Products, Inc.
35 Pond Park Road
Hingham, MA 02043
617/741-5008

Accepted filter models: Focus Plus FFGP31 Dark and FFGP62 Clear.

NEC Technologies, Inc.
1255 Michael Drive
Wood Dale, IL 60191
708/860-9500

Accepted filter models: MultiSync Monitor Lens. Standard tint.

Sunflex
1029 Corporation Way
Palo Alto, CA 94303
415/962-0488

Accepted filter models: Kristal Clear and OptiView.

Taiwan Union Glass Industrial Co.,
11th Floor, 26, Jen-ai Road, Sec. 3
Taipei, Taiwan R.O.C.
886 2 702 7126

Accepted filter models: UNUS AC-143, AC-145. AC-173, AC-175, IC-143, IC-145 and DC-203.

ACCO USA
770 South ACCO Plaza
Wheeling, IL 60090
708/541-9500

Accepted filter models: Glarecare GS and Glarecare VS.

Image One, Inc.
19 West 34th Street
New York, NY 10001
800/522-1231

Accepted filter models: Image One Glass and Image One Glass Plus.

Startech Computer Products
100 Piccadilly Street, Unit 103
London, ON N6A 1R8
CANADA
519/438-8529

Accepted filter models: Monitor Dock-it/AR14CF.

Appendix F

Additional Resources

American College of Occupational and
Environmental Medicine
55 W. Seegers Road
Arlington Heights, IL 60005
847-228-6850

American Industrial Hygiene Association
2700 Prosperity Avenue, Suite 250
Fairfax, VA 22031
703-849-8888

American National Standards Institute
1430 Broadway
New York, NY 10018
212-642-4900

American Optometric Association
243 North Lindbergh Blvd.
St. Louis, MO 63141
314-991-4100

American Society of Safety Engineers
1800 East Oakton Street
Des Plaines, IL 60018-2187
847-699-2929

American Society for Testing and
    Materials
1916 Race Street
Philadelphia, PA 19103
215-299-5400

Center for Office Technology
575 Eighth Avenue, 14th Floor
New York, NY 10018
212-560-1298

CTD News
LRP Publications
747 Dresher Road
PO Box 980
Horsham, PA 19044-0980
800-341-7864
215-784-9639 Fax

College of Optometrists in Vision
    Development
P O Box 285
Chula Vista, CA 91914
619-425-6191

Corporate Vision Consulting
The Eye-CEE System for VDT Users®
2404 Sacada Circle, Suite A
La Costa, CA 92009
800-383-1202 (voice/fax)
e-mail: eyedoc@adnc.com
http://www.netpagecom.com/CVC

Dept. of Health and Human Resources
200 Independence Ave., SW
Washington, DC 20201
202-619-0257
http://www.os.dhhs.gov

Department of Labor
200 Constitution Ave., NW
Washington, DC 20210
202-219-7316

Equal Employment Opportunity
    Commission (ADA)
1801 L Street, NW
Washington, DC 20507
202-663-4900

Human Factors and Ergonomic Society
P O Box 1369
Santa Monica, CA 90406-1369
310-394-1811

Illuminating Engineering Society
345 East 37th Street
New York, NY 10017
21-705-7916

National Institute for Occupational Safety
    and Health (NIOSH)
Centers for Disease Control
1600 Clifton Road, NE
Atlanta, GA 30333
404-639-3534
http://www.cdc.gov/niosh/homepage.html

National Safety Council
1121 Spring Lake Drive
Itasca, IL 60143-3201
708-285-1121

Optometric Extension Program
Foundation, Inc.
1921 E. Carnegie, Suite 3L

Santa Ana, CA 92705
714-250-8070

Occupational Safety and Health
 Administration (OSHA)
US Dept. of Labor
200 Constitution Avenue, NW
Washington, DC 20216
800-282-1048
http://www.osha.gov

Occupational Vision Service, Inc.
4100 Avenida de la Plata, Suite A
Oceanside, CA 92056
800-861-4684

Prevent Blindness America
500 E. Remington Road
Schaumburg, IL 60173
800-331-2020

Appendix G

Title 8

General Industry Safety Orders
Section 5110, Ergonomics

Readopted by the Occupational Safety and Health Standards Board on April 17, 1997
Approved by OAL on June 3, 1997
Effective July 3, 1997

Group 15. Occupational Noise and Ergonomics
Article 106. Ergonomics
Section 5110. Repetitive Motion Injuries

(a) Scope and application. This section shall apply to a job, process, or operation where a repetitive motion injury (RMI) has occurred to more than one employee under the following conditions:

(1) Work related causation. The repetitive motion injuries (RMIs) were predominantly caused (i.e. 50% or more) by a repetitive job, process, or operation;

(2) Relationship between RMIs at the workplace. The employees incurring the RMIs were performing a job process, or operation of identical work activity. Identical work activity means that the employees were performing the same repetitive motion task, such as but not limited to word processing, assembly, or loading;

(3) Medical requirements. The RMIs were musculoskeletal injuries that a licensed physician objectively identified and diagnosed; and

(4) Time requirements. The RMIs were reported by the employees to the employer in the last 12 months but not before July 3, 1997.

Exemption: Employers with 9 or fewer employees.

(b) Program designed to minimize RMIs. Every employer subject to this section shall establish and implement a program designed to minimize RMIs. The program shall include a worksite evaluation, control of exposures which have caused RMIs and training of employees.

(1) Worksite evaluation. Each job, process, or operation of identical work activity covered by this section or a representative number of such jobs, processes, or operations of identical work activities shall be evaluated for exposures which have caused RMIs.

(2) Control of exposures which have caused RMIs. Any exposures that caused RMIs shall, in a timely manner, be corrected or if not capable of being corrected have the exposures minimized to the extent feasible. The employer shall consider engineering controls, such as work station redesign, adjustable fixtures or tool redesign, and administrative controls, such as job rotation, work pacing or work breaks.

(3) Training. Employees shall be provided training that includes an explanation of:

(A) The employer's program;

(B) The exposures which have been associated with RMIs;

(C) The symptoms and consequences of injuries caused by repetitive motion;

(D) The importance of reporting symptoms and injuries to the employer; and

(E) Methods used by the employer to minimize RMIs.

(c) Satisfaction of an employer's obligation. Measures implemented by an employer under subsection (b)(1), (b)(2), or (b)(3) shall satisfy the employer's obligations under that respective subsection, unless it is shown that a measure known to but not taken by the employer is substantially certain to cause a greater reduction in such injuries and that this alternative measure would not impose additional unreasonable costs.

Note: Authority cited: Sections 142.3 and 6357, Labor Code. Reference: Sections 142.3 and 6357, Labor Code.

Glossary

Accommodation: In regard to the visual system, accommodation is the focusing ability of the eye.
Acuity: A measure of the ability of the eye to resolve fine detail, specifically to distinguish that two points separated in space are distinctly separate.
Astigmatism: A visual condition in which the light entering the eye is distorted such that it does not focus at one single point in space.

Behavioral optometry: A branch of optometry based on a model of vision which addresses a 'holistic' approach to visual function, stating that vision and visual abilities can be trained or enhanced.
Binocularity: The use of two eyes at the same time, where the usable visual areas of each eye overlap to produce a three-dimensional perception.
Brightness: The subjective attribute of light to which humans assign a label between very dim and very bright (brilliant). Brightness is perceived, not measured. Brightness is what is perceived when lumens fall on the rods and cones of the eye's retina. The sensitivity of the eye decreases as the magnitude of the light increases, and the rods and cones are sensitive to the luminous energy per unit of time (power) impinging on them.

Cataracts: A loss of clarity of the crystalline lens within the eye which causes partial or total blindness.
Cathode ray tube (CRT): A glass tube that forms part of most video display terminals. The tube generates a stream of electrons that strike the phosphor-coated display screen and cause light to be emitted. The light forms characters on the screen.
Color convergence: Alignment of the three electron beams in the CRT that generate the three primary screen colors – red, green and blue – used to form images on-screen. In a misconverged image, edges will have color fringes (for example, a white area might have a blue fringe on one side).
Color temperature: A way of measuring color accuracy. Adjusting a monitor's color-temperature control, for example, may change a bluish white to a whiter white.
Convergence: That visual function of realigning the eyes to attend an object closer than optical infinity. The visual axes of the eyes continually point closer to each other as the object of viewing gets closer to the viewer.

Diplopia (double vision): That visual condition where the person experiences two distinct images while looking at one object. This results from the breakdown of the coordination skills of the person.
Dot matrix: A pattern of dots that forms characters (text) or constructs a display image (graphics) on the VDT screen.

Dot pitch: The distance between two phosphor dots of the same color on the screen.

DRAM (Dynamic Random Access Memory): Pronounced 'dee-ram'. The readable/writable memory used to store data in personal computers. DRAM stores each bit of information in a cell composed of a capacitor and a transistor. Because the capacitor in a DRAM cell can hold a charge for only a few milliseconds, DRAM must be continually refreshed in order to retain its data.

Electromagnetic radiation: A form of energy resulting from electric and magnetic effects which travels as invisible waves.

Ergonomics: The study of the relationship between humans and their work. The goal of ergonomics is to increase worker comfort, productivity and safety.

Eyesight: The process of receiving light rays into the eyes and focusing them onto the retina for interpretation.

Eyestrain (asthenopia): Descriptive terms for symptoms of visual discomfort. Symptoms include burning, itching, tiredness, aching, watering, blurring, etc.

Farsightedness (hyperopia): A visual condition where objects at a distance are more easily focused, as opposed to objects up close.

Font: A complete set of characters including typeface, style, and size used for screen or printer displays.

Focal length: The distance from the eye to the viewed object needed to obtain clear focus.

Hertz (HZ): Cycles per seconds. Used to express the refresh rate of VDTs.

Holistic: An attempt to see the whole situation and to treat the person, not as individual parts, but as a whole performance system.

Illuminance: The luminous flux incident on a surface per unit area. The unit is the lux, or lumen per square meter. The foot-candle (fc), or lumen per square foot is also used. An illuminance photometer measures the luminous flux per unit area at the surface being illuminated without regard to the direction from which the light approaches the sensor.

Interlaced: An interlaced monitor scans the odd lines of an image first, followed by the even lines. This scanning method does not successfully eliminate flicker on computer screens.

Lag: In optometric terms, the measured difference between the viewed object and the actual focusing distance.

LCD (Liquid Crystal Display): A display technology that relies on polarizing filters and liquid-crystal cells rather than phosphors illuminated by electron beams to produce an on-screen image. To control the intensity of the red, green and blue dots that comprise pixels, an LCD's control circuitry applies varying charges to the liquid-crystal cells through which polarized light passes on its way to the screen.

Light: The radiant energy that is capable of exciting the retina and producing a visual sensation. The visible wavelengths of the electromagnetic spectrum extend from about 380 to 770 nm. The unit of light energy is the lumen.

Luminous flux: Visible power, or light energy per unit of time. It is measured in lumens. Since light is visible energy, the lumen refers only to visible power.
Luminous intensity: The luminous flux per solid angle emitted or reflected from a point. The unit of measure is the lumen per steradian, or candela (cd). (The steradian is the unit of measurement of a solid angle.)
Luminance: The luminous intensity per unit area projected in a given direction. The unit is the candela per square meter, which is still sometimes called a nit. The footlambert (fL) is also in common use. Luminance is the measurable quantity which most closely corresponds to brightness.

MHz (MegaHertz): A measurement of frequency in millions of cycles per second.
Mouse: A computer input device connected to the CPU.
Myopia (near-sightedness): The ability to see objects clearly only at a close distance.
Muscoskeletal: Relating to the muscles and skeleton of the human body.

Nearpoint: The nearest point of viewing, usually within arm's length.
Non-interlaced: A non-interlaced monitor scans the lines of an image sequentially, from top to bottom. This method provides less visible flicker than interlaced scanning.

Ocular motility: Relating to the movement abilities of the eyes.
On-screen controls: On-screen controls let you change settings as you would program a VCR. Visual feedback is provided on-screen as you push certain buttons.

Perception: The understanding of sensory input (vision, hearing, touch, etc.).
Pixel: The smallest element of a display screen that can be independently assigned color and intensity.
Phosphor: A substance that emits light when stimulated by electrons.
Presbyopia: A reduction in the ability to focus on near objects caused by the decreased flexibility in the lens, usually due to the age of the person being over 40 years old.
Polarity: The arrangement of the light and dark images on the screen. Normal polarity has light characters against a dark background; reverse polarity has dark characters against a light background.
Presets: Many monitors offer control presets that enable you to switch between different resolutions and color depths. The number of custom settings varies from monitor to monitor, and ranges from around 4 to 28 settings.

RAM (Random Access Memory): Pronounced 'ram'. The generic term for read/write memory – memory that permits bits and bytes to be written to it as well as read from it – used in modern computers.
Refractive: Having to do with the bending of light rays, usually in producing a sharp optical image.
Refresh rate: The number of times per second that the screen phosphors must be painted to maintain proper character display.

Resolution: The number of pixels, horizontally and vertically, that make up a screen image. The higher the resolution, the more detailed the image.

Resting Point of Accommodation (RPA): The point in space where the eyes naturally focus when at rest.

Suppression: The 'turning off' of the image of one eye by the brain, most often to avoid double vision or reduce excess stress.

Swim: A wave-like motion of screen display information, usually in a vertical direction, due to electrical malfunctioning in the VDT.

SVGA (Super Video Graphics Array): A video adapter capable of higher resolution pixels and/or colors than the $320 \times 200 \times 256$ and $640 \times 480 \times 16$ which IBM's VGA adapter is capable of producing. SVGA enables video adapters to support resolutions of 1024 by 768 pixels and higher with up to 16.7 million simultaneous colors (known as true color).

VDT (video display terminal): An electronic device consisting of a monitor unit (e.g. cathode ray tube), connection to a computer central processing unit and input device.

Vision: A learned awareness and perception of visual experiences (combined with any or all other senses) that results in mental or physical action. Not simply eyesight.

Visual stress: The inability of a person to visually process light information in a comfortable, efficient manner.

Vision therapy: A treatment (by behavioral optometrists) used to develop and enhance visual abilities.

VRAM (video RAM): Pronounced 'vee-ram'. A dual-ported RAM design that lets the memory chip read and write simultaneously. The higher throughput results in higher resolution and color modes. VRAM enables a Block Write feature, which is useful for handling video.

Index

accommodation 11–12, 18, 22, 40, 43, 45, 53, 85, 97, 107–10, 113
Americans with Disabilities Act 107
ANSI 27, 58–60, 74, 110–12
anti-reflective coating 83
aqueous fluid 6
asthenopia 38
astigmatism 9, 38–40, 98, 100, 110

backache 37
bandwidth 24, 33
bifocal 42, 80–3, 97, 105
binocular vision 11
blink rate 41, 100
Braille 66, 67
brightness 20–1, 26, 28–30, 32, 34, 44, 71, 73, 78–9

choroid 6
color 5–7, 21, 24, 26, 29–34, 37, 44, 46, 58, 61, 67, 70, 73, 78
Computer Vision Syndrome 37, 46, 98, 127
cones 7, 46
contact lenses 50, 60–1, 93, 97–8, 110
contrast 21, 26, 28–30, 32, 34, 53, 58, 67, 70–1, 78–9, 109
convergence insufficiency 38
cornea 5–6, 8, 9
critical fusion frequency (CFF) 30
CRT 23–4, 26, 81
Cumulative Trauma Disorders (CTDs) 16

diplopia 45
dot pitch 24
DPMS 32
DRAM 33

EC Directives 114
emmetropia 7
ergonomics 15–17, 22, 31, 37, 42, 46–7, 51, 65, 74, 101–4, 107, 110, 112–14, 116
eye exercises 84
eye injury 58
eyeglasses 80, 110
eyestrain 3, 18, 19, 32, 37, 38, 45, 71, 74, 78, 83, 85

farsightedness 8
FDA 60, 111
flicker 26, 29–33, 53, 74
Frankfurt Line 44

glare 11, 20–1, 28–9, 39–40, 44, 59, 69, 70–3, 83, 91, 96

headaches 9, 37–9, 74, 96

hyperopia 8, 38–40, 110

IES 78, 91, 112
iris 5–6, 11
ISO 28, 31, 112

jitter 26, 31

lag of accommodation 18, 40
LCD 23–4, 26
legal blindness 63
legibility 26–7, 29
light sensitivity 37, 39, 44
lighting 2–3, 17–18, 20–2, 28–9, 35, 39, 44, 51, 59, 67, 69, 72–4, 78–9, 83, 95, 97, 111–12
   Accent 20
   Ambient 20
   Task 20
low vision 63

McCullough Effect 46
myopia 7–8, 12–13, 18, 22, 38, 40, 46–7, 95, 110

Nearsightedness (see myopia)

Optical Character Recognition (OCR) 67
ophthalmologist 49–50, 52
optician 50
optometrist 37, 49–52, 91, 110
OSHA 58, 60, 102, 103, 111–14

photoreceptors 7
pixels 24
polycarbonate lenses 60
presbyopia 40, 97
progressive addition lens (PAL) 82, 97, 106

radiation 56, 59, 95–6
repetitive motion injuries 16, 102, 114
resolution 26, 35
   addressable 26
   image 26
Resting Point of Accommodation 18
retina 6–8, 11–13, 29–30, 39–40, 45–6, 87
rods 7

scanning 11
sclera 5–6
shadow mask 24
Snellen chart 10
standards 51, 57, 59–60, 62, 74, 96, 102, 107, 110–12, 114
suppression 11, 45, 95
tracking 11, 66, 85, 87
trifocal 81–2

vision screening 52, 106
vision therapy 38, 51, 84, 94
visual acuity 10–11, 28, 39, 51, 63, 75
visual stress 17, 71, 79, 94, 98
vitamins 97–9

vitreous humor 6–7
VRAM 33

World Wide Web 67–8